Adults count too

Mathematics for empowerment

Adults count too

Mathematics for empowerment

Roseanne Benn

For Cathy, Anna and Louise

Published by the National Institute of Adult Continuing Education
(England and Wales)
21 De Montfort Street, Leicester, LE1 7GE
Company registration no. 2603322
Charity registration no. 1002775

First published 1997

CATALOGUING IN PUBLICATION DATA
A CIP record for this title is available from the British Library
ISBN 1 86201 007 2

Typeset by Midlands Book Typesetting Co.
Text design by Virman Man
Cover design by Sue Storey
Reprinted 2004 (twice)

Contents

Acknowledgements

As this book is the culmination of so many years' work, I would like to thank all my students, colleagues and friends for their support over the years.

In particular, I should like to thank:

my father for his early interest and support in my mathematics education.

my students and colleagues at Kingston Polytechnic for introducing me to many cultures and many levels of mathematics and particularly my part-time mathematics degree students who taught me how to teach and the true meaning of the terms 'commitment' and 'perseverance'.

my students and colleagues at Hillcroft College in whose company I had both a great deal of fun and an introduction to notions of democracy and social justice and where I began my research in adult education, women's education, and the teaching and learning of mathematics in adult education.

my students and colleagues in the University of Exeter and throughout Devon and Cornwall in FE colleges and adult education centres. In their company, my development of Access provision and mathematics for Access brought together so much of my thinking. Some of these students were kind enough to say I changed their lives. They most certainly changed mine.

colleagues, nationally and internationally, who through conferences and other meetings unfailingly strengthened my resolve to write this book.

the UFC, subsequently HEFCE, for a research grant to support this work and Rob Burton who worked with me as my research assistant for two years.

my Department for study leave.

Jill Da Silva and Greta Tink for secretarial and other support.

Roger Fieldhouse for comments on earlier drafts and for his unfailing support.

Introduction

As a schoolgirl, I fell in love with mathematics. I thought it was a wonderful subject – to me, it epitomised elegance, beauty and simplicity. (It had the added advantage that there was minimal use of English and I was an abominable speller . . .)

My subsequent educational pathway was, as is the case for many of our adult students, very chequered. One small baby later, I embarked as a mature student on my mathematics degree. Like most adult learners, I was committed and highly motivated. After gaining my degree, I taught mathematics at the then Kingston Polytechnic to all kinds of learners including adults studying for a mathematics degree parttime, and all manner of mathematics service teaching. After two more children, I decided to move into continuing education to support women and men who, for whatever reason, had missed out the first time round and wished to return to study. I became more and more interested in opening up routes for adults to both mathematics and higher education. This career choice led me to Hillcroft College then later to the Department of Continuing and Adult Education at the University of Exeter.

Earlier in my career, I taught in a humanistic student-centred way but with a belief in the truth, beauty and abstraction of mathematics – a belief that mathematics is a 'peek into the mind of God', is already 'written in the sands'.

Three factors caused me to question this value-free, culture-free notion of mathematics. Firstly, through feminism, I became aware of the deeply unequal nature of both society and the constructs of society. Secondly, through my work in teaching, development and research over many years with adult returners, I became interested in the reasons that mathematics held the particular and peculiar place that it unquestioningly does in so many adults' minds. Lastly, but most importantly, I became committed to the notion that adult education has a vital role to play in a democratic society. I became convinced that the low levels of numeracy in our society limit participation and critical citizenship.

This book is therefore the result of all my working life – my teaching in mathematics itself, my work with adult returners, my research and my abiding interest in critical citizenship and the role mathematics has to play in this. It is my hope that it will be of interest to anyone concerned with mathematics, mathematics education, adult education and a democratic and more just society. It is written for people involved in the teaching and learning of mathematics but will also be of interest to all involved in adult education.

Adults count too examines the low level of numeracy in our society, the reasons why this is critical and the forces acting on adults which contribute to this state of affairs. These forces include experiences and philosophies of mathematics and mathematics education, social and cultural factors, political imperatives and the aims and aspirations of the adults who, despite all odds, wish to return to study to learn mathematics.

As theory is an essential basis for the acquisition of new ideas with consequent change in practice, the book examines these issues from a theoretical perspective. But it is also grounded in practice and the belief that adults learn best when mathematics teaching builds on positive attitudes, is interactive and cooperative, practical and relevant, set in a social, historical and cultural context and enjoyable and fun. It was written in the hope that it would contribute to a more empowering curriculum for adults learning mathematics, a curriculum which would, first and foremost, fulfil the priorities of the individual but would also take into account the different needs of the diverse population of adults learning mathematics, building on their life, work and social experience.

It recognises but rejects the discourse of mathematics for social control where mathematical literacy is seen as a way of maintaining the *status quo* and producing conformist and economically productive citizens. Similarly, it rejects the approach of deficit and disadvantage where any problem with mathematics is located in the learner rather than the system. It moves away from an individual skillsbased approach to one of a critical analysis of social and economic factors, a cultural critique of the elitist assumptions about mathematics and mathematics education and a critique of the educational system with its tendency to create deficit models for adults returning to learn mathematics. It is based on the passionate belief that mathematics is a crucial way of knowing that can make a real difference in people's lives but only if it can be seen in the wider context of society, structural inequality and cultural difference.

Our modern society has been dominated by certain cultures and ways of thinking. In the new postmodern world, this is being questioned at all levels. There is a growing recognition that there are alternative world views, truths, realities and cultures, many of which are not recognised or valued by society. This is true in mathematics and mathematics education and this book is concerned with the discovery and recognition of these 'silent' voices. Such an approach built on social, economic, political and cultural awareness, a value system of social justice and equity and a collaborative, cooperative approach to learning, might transform mathematics education for adults into education for empowerment.

How to use this book

The book can be read sequentially but the reader may also read selectively.

Section 1: Adult education

This section presents an overview of the adult education environment within which adults learn mathematics. Chapter 1 explores the role and purposes of adult education in our society and Chapter 2 identifies some characteristics of and influences on participants in the process as either learners or tutors.

Section 2: The framework within which adults learn mathematics

Readers familiar with the adult education environment or who wish to get straight to issues involving mathematics can start with this section which outlines the wider disciplinary, social, political and cultural context in which adults learn mathematics. Chapter 3 examines underlying beliefs about the nature of the subject itself, counterposing an absolutist philosophy of mathematics with an alternative fallibilist approach. Chapter 4 introduces the three main players – the learner, the tutor and the curriculum – and identifies a matrix of forces which act on all three. In Chapters 5, 6 and 7 the players are positioned in the framework by an examination of both their aims and goals and the effect on learning and teaching approaches of current cultural, political and educational forces and experiences or philosophies of mathematics.

Section 3: Understanding adults learning mathematics

This section looks further at the factors affecting adults learning mathematics. Chapter 8 exposes the low level of adults' confidence and knowledge of mathematics whilst Chapter 9 looks at whether this is fundamentally injurious to a democratic society. Chapter 10 introduces the concept of discourse and how this can be utilised to facilitate the teaching of mathematics to adults. The next chapters (Chapters 11, 12, 13 and 14) concentrate on the importance of recognising and valuing diversity and difference through the notion of culture. Chapter 15 builds on this by illustrating how individuals acquire an often very effective mathematics-which-works located in their own lives and culture but have difficulty transferring this knowledge into formal academic mathematics.

Section 4: Implications for practice

Chapter 16 builds on increased awareness of the framework in which adults learn mathematics to move to developing a more emancipatory curriculum.

Section 1

Adult education

Chapter One

The role and purpose of adult education

This chapter will attempt the difficult but interesting task of exploring the role and purpose of adult education in Britain in order to provide the educational framework within which adults learn mathematics in a formal setting. It attempts a critical consideration of the educational purposes of adult education by an examination of the social nature of the work, whose interests it serves and its place in the wider education system. After a brief discussion about *what* adult education is, it then moves to the *why* question. Why educate adults? Is it for the conservation and reproduction of society or are the underlying purposes be more radical? This is developed into an exploration of the issues around the present very real dichotomy between individualism and collectivism, then extended to an examination of the outcomes of the current preoccupation of adult education with personal development, student-centred learning and transformative education.

Adult education practice

Adult education has always been a difficult area of work to classify with adult educators teaching a wide range of subjects in a wide range of different circumstances to a wide range of students. As Steele suggests (1993: 35), adult educators range promiscuously over pedagogical, historical, political, psychological and practical issues. Adult education in Britain has traditionally been pluralist, with major strands provided by the local authorities, the voluntary organisations and the universities. This tradition of 'night school', of adult education providing a wide range of non-vocational, vocational and recreational courses at low cost in local centres, is disappearing to be replaced by 'continuing education' with its more utilitarian connotations. The emphasis in the 1944 Act on a partnership of providers collectively securing an adequate provision of further and higher education for all who are able and willing to benefit by it was largely displaced by the 1991 Education White Papers (DES 1991) and the 1992 Higher and Further Education Act. The move now is to more individualistic self-help values and a greater competitiveness between providers. The new formation is likely to be driven by ideological forces, through funding mechanisms, into mainstream higher education and mainstream further education.

But what is adult education? Deliberate, intentional and planned adult education has been categorised, if not defined, in the following way. Formal adult education is that provided by the education and training system set up or sponsored by the state for the express purpose. In Britain, it encompasses vocational education and training, human resource development, preparation for formal study, adult basic education, English as a second language, community education and liberal adult education. Non-formal adult education comprises the many deliberate educational enterprises set up outside the education system (for example, by other departments) or by agencies where education is a subordinate objective (for example, churches and trade unions). Informal adult education is that vast area of social transactions in which people are deliberately informing, persuading, telling, advising and instructing each other and deliberately seeking out information, advice, instruction, wisdom and enlightenment (Groombridge 1983). As adult educators, we need to remember that what takes place in the 'classroom', no matter how loosely defined, is just the tip of the iceberg. Much of what will be considered in this book will be formal adult education, nevertheless later chapters will acknowledge the crucial importance of not just the planned but also the incidental learning that takes place informally in, for example, the workplace or the home.

Adult education in Britain has been, and still is, undergoing fundamental changes. The main providers of formal adult education in Britain today are the local education authorities (LEAs), further education colleges, the universities and the Workers Educational Association (WEA). The LEAs have traditionally been a major provider with an eclectic offering of courses for leisure, domestic training, school qualifications, job skills and non-vocational, noncredit educational activities. The gradual erosion of the power of local authorities by central government climaxed for adult education in the 1992 Further and Higher Education Act. By limiting state funding to Schedule 2 courses (ie, those leading to certain qualifications, basic education and certain types of access and special needs provision), the government through the Further Education Funding Council almost eliminated the public subsidy of non-vocational and non-accredited courses in local authority provision.

A major (and until the 1990s almost hidden) provider of adult education is the further education sector. This was administered by local government and traditionally catered for that part of the population, the majority from the working class backgrounds, who did not go into higher education but sought vocational education, sub-degree level academic qualifications and compensatory provision such as literacy, numeracy and language skills. Many also provided substantial programmes of non-vocational adult and community education. In the 1980s, provision increased for the unemployed, women who wished to return to the job market and Access provision for those who wished to enter higher education as mature student. The majority of students in further education were and still are adults, many studying part-time. The 1992 Further and Higher Education Act was critical for adult education provision. It established colleges as corporations with the consequent rise of the new managerialism, increased competitiveness, financial rather than educational missions and corporate survival. The need to increase student numbers has encouraged colleges to turn to the adult market but again Schedule 2 has limited the range and purpose of provision.

Towards a learning society

The clarion call 'towards a learning society' permeates our lives in the 1990s and appears to have universal support. After all, who would not wish to work for such a commendable aim? But adult educators need to look beneath the rhetoric and ask what this slogan actually means in today's society. They have never been members of a single society, but rather part of a heterogeneous series of overlapping and inter-related local, regional, national, international, global societies. In these 'globalised conditions', current conceptions of a learning society are dominated by the notion of the 'learning market' which emphasises support for the competitiveness of the economy as the purpose of lifelong learning. This is at the expense of other aspects of the learning society which include the learning society as 'an educated society', committed to active citizenship, liberal democracy and equal opportunities; and as 'learning networks', in which learners adopt a learning approach to life, drawing on a wide range of resources to enable them to develop their interests and identities (Edwards 1995). Before lining up behind the learning society banner, adult educators should investigate what is on it: the 'why' factor. Why embark on or encourage or promote this perception of the learning society?

Before continuing this discussion, it is perhaps worth recording the practical reality of much adult education. This was explored by Duke (1990) in a paper on the connections, if any, between adult education, social movements and democracy and bears repeating. Most adult education in Britain is to do with usually decent people being paid usually small amounts of money to teach other usually decent people some knowledge or skills which they want or need to acquire either to survive or do better or be happier and more fulfilled in their daily lives. Many tutors have a commitment to helping those in society who are disadvantaged for whatever reason. Some tutors have in addition an overt or covert agenda to work through the medium of adult education towards social justice and a more equal society by the raised awareness or (ugly but useful word) the conscientisation of their students. A few become activists with their students. Many, probably most, see adult education as a politically and ideologically uncommitted form of activity intended solely to promote individuals' learning to different specific ends. Adult education as a profession and a form of provision is thus largely perceived as neutral. It is probably reasonable to say that not many of the large army of part-time tutors and the large number of full-time tutors would consider adult education as a social movement. Nevertheless, there is a body of thought that states that no education is neutral, that to have no purpose is to have the purpose of the continuation of the *status quo* and that all activities are therefore political. These contradictory views are interesting enough to warrant further discussion.

The ideological continuum

Within adult education, there is a continuum of attitudes ranging from the conflict view of society through to the consensus view of society.

> The concepts of adult education as either a radical force, primarily seeking to make a major impact upon society, or as a conservative force devoted to the atomistic view of man (*sic*) are, in everyday life, subject to considerable constraints. These concepts are, therefore, best seen as part of a continuum. At one extreme, we are dealing with the view that any system of adult education must, if it is to be effective, challenge established economic, political and social assumptions. At the other, we encounter the argument that adult education should only be concerned with the conservation of traditionally accepted normative standards. Between these two positions, there are a number of less extreme views which recognise that while there may be inevitable social and political concomitants of all educational activities, there is a spectrum of legitimate educational aims in which some are closer to the management of social change and others closer to the conservation of inherited cultural traditions.
>
> (Thomas and Harries-Jenkins 1975: 112)

The most influential parts of this continuum will now be examined in more depth.

Education for conservation and reproduction

The predominant function of all state-provided education is to conserve, reproduce and perpetuate society, to socialise people into society and to equip them to contribute to society in required ways. Even in societies where culture, economy and technology are changing rapidly, governments still wish to contain and conserve. This creates a tension between the need to preserve society and the need to prepare people for an unknown and uncertain future which is often dealt with by stressing technological change and the consequent need for new knowledge and skills whilst emphasising cultural continuity by, for example, encouraging patriotism or even nationalism.

Adult education, as part of the system, must contribute to the reproduction of social structures and power relationships by concentrating on the socialisation of individuals into society. This it does by its emphasis on work-related courses and second chance education, both forms of provision which encourage the incorporation of individuals into society by giving them a stake in that society.

Foucault (1986) extends these ideas in his thoughts on power and knowledge as expressed in his discussions on 'disciplines'. He uses the term 'discipline' both as the delineation of bodies of knowledge but also to replace terms like 'profession' as in 'teaching profession'. This allows the term 'discipline' to capture aspects of power that are normally masked. He links the subject area and its conceptual structure to the notion of discipline meaning subjection and obedience arguing that formal education exercises power through the organisation of space, time and capacities. This is particularly true of initial education but also applies to adult education. Activities are carefully organised not just with rooms and timetables controlling distribution and rhythms but also through the development of the curriculum. What activities are appropriate to any particular stage will depend essentially on that discipline's true discourse – that is the knowledge that has been established through the exercise of power within that disciplinary block. Examinations, classifications and remedial treatment establish normal patterns of expectations. This knowledge, developed *through* the exercise of power, is used *in* the exercise of power to produce normalised individuals. This is the link between discipline as cognitive structures and as submission or subjection. In essence, a discipline is a body of knowledge with a system of social control (Marshall 1989).

Given this framework of power/knowledge, shifts from traditional didactic teacher-centred approach in adult education are not an abandonment of the exercise in power over the student but rather a change in the technologies and programmes of power. Power is still exercised in the search for normalised and governable people. If it is more humane, more subtle and less overt it may also be more dangerous. In person-centred pedagogy, discipline is shifted from visible to invisible, overt to covert regulation. Writers such as Walkerdine (1988) argue that this is all at one with current attempts to govern through bourgeois democracy. The 'caring professions', including adult educators, are harnessed by various means including financial to render the governed governable. Issues of power/knowledge, control and discipline are arguably central in any discussion of education, adult education or the education of adults who wish to learn mathematics.

The liberal tradition

This tradition inherits a broad liberal ideology from the university extension movement of the 19th century and encompasses notions of individual self-fulfillment, social purpose, public service, social justice and class emancipation (Fieldhouse 1985). Objectivity and political neutrality are essential elements of this tradition. A central tenet is that students and tutors pursue their studies together in an atmosphere of open enquiry, free of all prejudice and political propaganda, considering all possible answers to their enquiries before ultimately alighting upon the truth. This method of teaching endeavours to

- develop students' critical faculties to enable them to acquire a critical attitude to all judgements about the objects studied, including and especially their own judgements;

- to distinguish as carefully as possible between matters of fact and matters of interpretation;
- to ensure that all basic assumptions, both those held by the tutor and those held by the students, are questioned;
- to encourage students to distinguish between significant questions, which can open up fruitful lines of investigation, and trivial questions;
- to accustom students to understand, state and criticise theories and points of view which they do not themselves accept.

Five central criteria of a liberal education have been suggested:

- the development of individual autonomy and hence individual responsibility for the consequences of one's actions;
- individual understanding about the universe, one's place within it and one's interaction with it;
- individual freedom;
- impartiality and objectivity;
- crucially, respect for persons as unique and autonomous centres of consciousness and action.

These criteria presuppose the possibility of fallibility and hence the plurality of claims of knowledge (Bagnall 1991). Although mathematical education is not usually perceived as liberal, later chapters will be concerned with many of the ideas and criteria discussed above including the notions of fallibility and plurality.

The radical tradition

The fundamental tenet of radical adult education is that education can never be neutral, that it always operates in a political, economic, cultural and social environment and that having no social purpose is in fact to have the social purpose of the continuation of the *status quo*. Radical educationalists, claiming to represent the disenfranchised and seeking to build a better order, have always wished to bring about social change through education and, in the past, this has focused on the development of citizenship and the extension of participation in the social democratic political purpose. The philosophies of radical adult education have emphasised its orientation towards the working class and the economically and politically disenfranchised. The aim has been to provide individuals with the knowledge which they can use collectively to change society if they so wish, and to equip members of the working class with the intellectual tools to play a full role in a democratic society or to challenge the inequalities and injustices of society in order to bring about social change (Fieldhouse 1985: 2). Again these notions will be central to later discussions on mathematics and active participation in society.

The new realism

The 'New Realism' of the New Right places an instrumental purpose upon the role of education. The 1987 White Paper *Higher Education: Meeting the Challenge* set out the agenda for higher education and confirmed its focus upon instrumentalism by, for example, defining access to be 'taking account of the country's need for highly qualified manpower' (DES 1987). The 1992 FHE Act set out clearly, through Schedule 2, those areas that the Government saw as national priorities. As already stated, these are accredited courses, access to accredited courses and basic skills. This new focus upon vocationalism has received considerable financial support through the Further Education Funding Council (FEFC) together with organisations such as the Training, Enterprise and Education Directorate (TEED) and local Technical and Enterprise Councils (TECs) (Benn and Fieldhouse 1994; Field and Weller 1993). Consequently, it can be argued, the mainstreaming of the vocationalist thrust of the recent past has shifted the focus of attention away from any sense of social purpose in adult education to an image conceived of in primarily institutional and functional terms. One of the categories of provision explicitly mentioned in Schedule 2 are courses to teach the basic principles of mathematics. It is a subject clearly identified as underpinning the prosperity of Britain plc.

The functionalist approach

Looking at these same issues from a different angle, it can be argued that functionally, education has three roles. The first is a clear technical role in providing society with a trained workforce. Secondly, education reproduces society: the individual is socialised into the dominant social and cultural norms. Thirdly, in a stratified society that has allegedly moved on from being a class-based society to a more meritocratically-based system, education establishes status. Social roles and positions are set by the level of education received. Progression through merit is to the more prestigious institutions of learning and higher level employment.

Adult education fits into this functionalist approach to education to a certain degree as has been shown by earlier examination of the first and second points. We will now turn our attention to the third point. In modern societies, qualifications are used to legitimise inequalities of pay and status and this leads to limitation of access as parents endeavour to pass their own class advantage on through educational opportunities to their children. Cross-national comparisons of education show that determinants of access still partly lie in cultural, social and economic factors (Halsey 1992). The cultural heritage transmitted through the school system may be offered to all children equally but is not received equally because children are 'already the recipients of social and cultural inequalities transmitted through the family' (Westwood 1981: 39). By treating all pupils, however unequal they may be in reality, as equal in rights and duties, the educational system is led to give its *de facto* sanction to initial cultural inequalities (Bourdieu 1971).

This notion of structural inequality needs to be examined in the context of adult education (Payne 1992). Adult educators are used to developing pedagogy to meet student needs rather than within a restricted traditional form. The whole basis of their practice is grounded in the use of appropriate teaching and learning styles. While adult education courses are still primarily built around individual learning, many adult educators recognise the structural causes of educational disadvantage, noting that, for example, people from the working class, women wishing to study science and technological subjects and people from different cultural backgrounds may have experienced a structural hindrance to their education.

The practice and policy that results from this tension between individual and structural experiences of education has created some of the collective experiences of adult education. On the one hand there is the institutional framework which judges success through individual hard work and attainment. A course succeeds when the students successfully complete it as individuals; thus the whole thrust of the pedagogy must result in individual success. However, the experience the students have of themselves as failures is created through the structures of inequality. Student-centred education must take account of this by 'starting where the student is'. For the students' educational potential to be unfrozen, their individual experience of education must be confronted alongside the structural experience of inequality (Corrigan 1992). This is as true for mathematics education as for any other.

Dominant ideas in current adult education practice

Having examined, if only briefly, some of the traditional purposes of adult education, we will now conclude this chapter with an exploration of some of the concepts or theories which, whether knowingly and consciously or otherwise, are influential in current adult educational practice. We will start with current interest and commitment to self-directed, experiential, problem-solving, learner-centered practice which has its roots in the powerful combination of humanism and behaviourism and which finds expression in Knowles' theory (or rather practice) of andragogy, the technology of the education of adults. This approach is characterised by four assumptions about adult learners:

- adults are self-directed learners;
- they have accumulated experience which is a rich learning source;
- they are ready to learn when the learning will help them cope with the tasks and problems of their life;
- they need practical competence rather than theoretical knowledge

(Knowles 1980, 1984).

If there is any one feature which is consistently held up as the identifying characteristic of adult education, it is the ability, and indeed the propensity, of adult learners to be self-directed (Candy 1981). They 'take the initiative, with or without the help of others, in diagnosing their learning needs, formulating learning goals, identifying human and material sources for learning, choosing and implementing appropriate learning strategies, and evaluating learning outcomes' (Knowles 1975: 18). Knowles (1972) also notes that 'as an individual matures he accumulates an expanding reservoir of experience that causes him to become an increasingly rich resource for learning, and at the same time provides him with a broadening base to which to relate new learnings (*sic*)'. These arguments have been used by Knowles and others to change the emphasis in adult education from the transmittal techniques of traditional teaching (pedagogy) to experiential techniques (andragogy).

Knowles' theory, or more appropriately technology, has been widely criticised as being inadequate, limited and limiting. (Collins' thoughtful critique (1991) shows clearly how these ideas have dominated practice in North America to a far greater extent than in Britain.) Nevertheless, this concept of the adult as a self-directing personality has had two major influences in the education of adults (Percy 1995). One was the impact on teachers in institutions, many of whom adopted the strategies of student-centred learning with great enthusiasm. The other led to investigations round whether adults as self-directing personalities might undertake learning away from formal learning settings. Tough's work (1971) showed not only the considerable amounts of self-directed study undertaken (a median of eight learning projects a year with 700 hours spent a year on learning projects being common) but also that much of this takes place outside the formal education system. The little research in this area that has been carried out in Britain (Brookfield 1980, Percy *et al* 1988, Sargant 1990, and Strong 1977) indicates that adult self-directed learning external to formally provided programmes is common in Britain but is often not identified by the learners themselves as learning. It regularly takes place in the home; and is frequently work-related. These findings that adults instigate their own 'unlabeled' learning will be echoed throughout this book.

There are, however, well-rehearsed arguments (see Collins 1991, Fieldhouse 1993b), Brookfield 1989) that self-directed learning, whether on a taught course or in independent study, has major limitations. This approach can be too comfortable, can allow existing assumptions and prejudices to go unchallenged and leave untouched aspects of the learner's values and actions that they would prefer not to examine. It can also be argued that a programme of self-directed independent study, while meeting the individual's perceived needs, may preclude the consideration of critical social, cultural and political influences and therefore excludes social benefit and the collective good from any assessment of value (Fieldhouse 1993b: 242). This mode of study may lack objectivity, allowing the adult educator's practice to remain subjective and normative. As Collins argues (1991: xi), in the absence of a sustained social critique, the learner becomes vulnerable to dominant pressures or ideologies.

But the concept of the adult as a self-directed learner is a powerful one which has been developed in a more rigorous way by Mezirow (1981, 1994).

His ideas locate learning in a wider social economic and political framework, are compatible with liberal ideas and have some common ground with the best of competency-based theories. Taking learning to be the social process of constructing and appropriating a new or revised interpretation of the meaning of one's experiences as a guide to action, Mezirow constructed a critical theory of adult learning. He found, through his empirical research with women participating in re-entry programmes, that the most distinctly adult domain of learning was perspective transformation. This is the emancipatory process of becoming critically aware of how and why the structure of psycho-cultural assumptions has come to constrain the way we see ourselves and our relationships, reconstituting this structure to permit a more inclusive and discriminating integration of experience and acting upon these new understandings. It is the learning process by which adults come to recognise their culturally-induced dependency roles and relationships, and the reasons for them, and to take action to overcome them. The most significant learning involves critical reflection. For this kind of learning, he identified the following phases:

- a disorienting dilemma;
- self-examination with feelings of guilt or shame;
- a critical assessment of assumptions;
- recognising that one's problems are shared and not exclusively a private matter;
- exploring options for new roles, relationships and ways of action;
- planning a course of action;
- acquiring knowledge and skills for implementing one's plans;
- provisionally trying out new roles;
- renegotiating relationships and negotiating new relationships;
- building competence and self-confidence in new roles and relationships and a reintegration into one's life on the basis of conditions dictated by one's new perspective

(Mezirow 1981: 126).

Mezirow identifies the prime role for adult educators as assisting adults to learn in a way that enhances their capability to function as self-directed learners where a self-directed learner is defined as having access to alternative perspectives for understanding his or her situation and for giving meaning and direction to his or her life, has acquired sensitivity and competence in social interaction and has the skills and competences required to master the productive tasks associated with controlling and manipulating the environment. Mezirow's *Charter for Andragogy*, although written in the technicist style so popular in North America and so deeply unpopular in Britain, is still useful reading for adult education practitioners no matter what discipline. The goals are to:

- progressively decrease the learner's dependency on the educator;

- assist the learner to define his/her learning needs – both in terms of immediate awareness and of understanding the cultural and psychological assumptions influencing his/her perceptions of need;
- assist learners to assume increasing responsibility for defining their learning objectives, planning their own learning programme and evaluating their progress;
- organise what is to be learnt in relationship to his/her current personal problems, concerns and levels of understanding;
- foster learner decision-making:
- facilitate taking the perspective of others who have alternative ways of understanding;
- encourage the use of criteria for judging which are increasingly inclusive and differentiating in awareness, self-reflexive and integrative of experience;
- foster a self-corrective reflexive approach to learning;
- facilitate problem-posing and problem-solving, including problems associated with the implementation of individual and collective action;
- help the recognition of relationships between personal problems and public issues;
- reinforce the self-concept of the learner as learner and doer by providing for progressive mastery;
- provide a supportive climate with feedback to encourage provisional efforts to change and to take risks;
- avoid competitive judgement of performance;
- make appropriate use of mutual support groups;
- emphasise experiential, participative and projective instructional methods;
- make appropriate use of modeling and learning contracts;
- make the moral distinction between (a) helping the learner understand his/her full range of choices and how to improve the quality of choosing versus (b) encouraging the learner to make a specific choice

(Mezirow 1981: 137).

Mezirow writes from a clearly articulated standpoint. He asserts that:

> an adult educator cannot be neutral in his or her conviction that social change is necessary to create a society in which all adult learners may participate fully and freely in critical reflective discourse. This is the necessary condition for adults to optimally participate in discourse to make meaning of their experience. As citizens, educators should become partisan activists to work towards creating such a society. As educators, we have an ethical commitment to help learners learn how to think for themselves rather than to consciously strive to convert them to our views. This commitment forbids us to indulge in indoctrination. What we can do is foster learner awareness of the need for change through transformative learning
>
> (Mezirow 1994: 230).

Not surprisingly Mezirow has been taken to task over these convictions. Tennant (1994), for example, whilst personally sharing much of this belief system, argues that adult educators need to distinguish clearly between normative and fundamentally transformative development. The former implies development and progress within a taken-for-granted world view and the latter results from the exposure and deconstruction of a given world view and its replacement by a new world view. He recognises the pragmatic reality mentioned earlier that many adult educators are explicitly concerned with providing educational support for normative life changes – those encountered within a rapidly-changing society with changing expectations and circumstances. Other adult educators are concerned with effecting more fundamental development transformations in programmes for example with themes of racism, sexism, poverty, illiteracy and unemployment. Innumeracy may well own a place in this list. Tennant is basically arguing that education can simply help learners to adjust to socially-expected development tasks (Freire's banking concept) or assist them to question fundamentally their perspectives on the world and their place in it (emancipatory concept). Adult educators need to make a distinction between these two educational outcomes.

This is not always easy. For example, a central feature of the discourse of adult education has been the concept of empowerment. There is a real dichotomy between the essentially liberal humanistic perspective, which declares that empowerment starts from within the person and is primarily about personal change, and a more critical and structural analysis that recognises the limitations and wishful thinking of a personal empowerment which does not relate to wider community and societal inequalities.

Claims have been made for adult education in raising working class consciousness, in combating social inequality and in developing collective empowerment when the reality has often fallen short of this. The danger is that adult education is only liberating and empowering at a very individual and restricted level and that it impacts in a very marginal way on local communities, society or the formal education system. To make a difficult task even harder, the very meaning of powerful words such as 'empowerment' are shifting. As Johnston (1993) points out, whereas at one time empowerment was understood to involve an emancipatory process which is in contrast to the notions of knowledge as a commodity on the one hand and as personal enlightenment on the other, now, in the nineties, it has been at the root of Major's Citizens Charter and central to the new consumer culture. So words which have traditionally reflected the social concerns of adult educators such as empowerment have been redefined and reduced to individual self-interest. The word 'entitlement', as another example, was recently used at a large gathering of adult educators to discuss the length of study in higher education with no reference to entitlement of access to the groups still under-represented. This colonisation of the vocabulary of adult education is particularly pernicious if it inhibits meaningful communication not just between adult educators and students but also within the body of adult educators themselves.

Having explored necessarily briefly the diverse roles and purposes of adult education, Chapter 2 will conclude the scene-setting by looking at the *who* questions in adult education: who are the students and who are the tutors?

Chapter Two

Some characteristics of and influences on adult learners and adult educators

In order to be able to understand more fully the various issues involved in adults learning mathematics, this chapter will extend our overall framework of adult education by examining the two crucial groups involved in the formal system: the students and the tutors.

Participation in adult education

The National Institute of Adult Continuing Education (NIACE) carried out a comprehensive review of the literature on participation and non-participation in formal adult education (McGivney 1990). Whilst noting that the whole concept of participation in such a large, diverse and complex sector is highly problematic and that theories of participation are woefully inadequate, McGivney was able to suggest the main attributes associated with adult participation. Participation obeys the iron law of education: the more education people have had, the more likely they are to want more and the more competent they will be at getting it (Wiltshire 1987). Age is a significant factor, arguably more so than gender or race, and young adults are more likely to engage in learning than those over fifty (Anderson and Darkenwald 1979).

Participants are characterised by certain favourable attributes, namely: higher income and occupational levels; the ability to anticipate and instigate social change; a higher level of schooling; and extended social relationships and cultural practices. Interestingly, educational participation is strongly connected with the extent of an individual's integration into community life. Participants are generally an active social minority leading a more diverse and intense social life and tend to be more involved in voluntary groups, political parties, unions, churches and local cultural activities. Whereas participants and non-participants engage equally in mass culture such as newspapers, television and holidays, participants and their families are significantly more engaged in cultural practices such as reading and visiting cinemas, theatres, museums and exhibitions. Hence participation in adult education arises from particularly tenacious social differentiations (Hedoux 1982).

McGivney found that the evidence suggested that irrespective of location or educational setting, certain sections of the community tend not to engage in any formal educational activity after leaving school – older adults; less well-educated people in lower social, economic and occupational strata; women with dependant children; minority ethnic groups and people living in rural areas. She asserts that non-participation is an indictment not of public apathy but of an education system which still projects a narrow and elitist image and which is complex, fragmented and diverse. Whilst situational barriers (such as time and cost) and institutional barriers affect participation, she argues that the major barriers to participation are attitudes, perceptions and expectations.

In summary, the literature on participation indicates that most people consider 'education is for other people'. It is difficult to establish precise figures but most research would indicate that less than 15 per cent of adults in Britain are participating in education at any one time and over half the population has not participated in learning in a formal setting since leaving school.

Are the non-participants the 'problem' or has the system itself failed to attract a large proportion of the population? The answer may partly lie in the importance adult educators attach to self-direction noted in the last chapter and the proud claim of many adult educators to meet individual needs by the promotion of personal growth through a student-centred curriculum. Unequal access will be inevitable in any service that operates from an assumption that all adults are self-directed and in circumstances which favour participation.

Even if the system, recognising social disadvantage, reacts with provision attempting to help individuals to adapt to dominant social and cultural norms, then alienation, non-participation or failure is likely to result. As will be shown later in this chapter, adult educators are well-educated: they have themselves been socialised into the norms of the dominant culture. It may well be that adult education attracts the students it does because it conforms to their notions about the nature and purpose of education. There is no outcry from those who have different notions because demand in this service is more often passive than active and voting with one's feet more common than attempts to change the situation. That adult education is based on a middle class and ethnocentric curriculum is evidenced by its participants. It is reasonable to ask if the individualistic and student-centred ideology of adult education supports some cultural styles but not others. As Keddie (1981) suggests, it is not a question as to whether individuals have needs or whether they should be satisfied but rather how those needs are socially and politically constructed, how they are articulated and whose voice is heard.

When adults do participate, the following characteristics may usefully describe some of their common features. Adults bring to their studies considerable knowledge and experience gained over the years and much of this 'knowledge-from-practice' will have been gained through trial and error, and learning from and with peers. Adults are autonomous learners who are used to setting their own goals, activities and timetables and will come with established attitudes and ways of doing things. They are likely to lack confidence in themselves as learners, under-estimate their own powers and be very worried about failure and looking foolish. This will be particularly true for those who have had poor experiences of learning in school or/and (more likely 'and') lifelong experiences of inequality and prejudice.

Adults tend to study part-time and have many other demands on their time so they seek relevance and immediate success in their goals. Relationship with the tutor is critically important in adult learning. Adults expect tutors to have a firm grasp of the subject matter, be enthusiastic, practise what they preach, and manage the overall learning situation effectively. They expect to work hard and be stretched. They expect constructive feedback, to enjoy and be actively engaged in the learning experience and to be treated with respect. One of the crucial differences between adult education and the education of the 'young' is that normally adults do not have to be there. If these expectation are not fullfilled they have the ultimate sanction of voting with their feet (Daines, Daines and Graham 1992).

Who learns mathematics and why

An International Seminar on Adult Numeracy (CUFCO 1993) which took place in France in 1993 with representation from throughout Europe, Latin America, the West Indies, Africa and Asia found a wide variation between adult learners of age, sex, mother tongue, professional situation and experience. However, a common characteristic of adults learning basic mathematics was their social background. Most belonged 'to the fringe of society' (p.58) and the report noted the weak or limited cultural fund most possessed. Most had limited initial education, often interrupted, with a high rate of failure. They rejected any links to this negative experience, with little self-confidence in their learning capabilities. Many were unemployed, with close relations also unemployed. Motivations for attending the course varied: finding a job, helping children to do their school work, better management of the household budget, managing a small family business, getting promotion at work, and so on.

This shows some contradiction between the whole body of adult learners and those learning mathematics. This may in part be due to methodological differences in data collection but may also reflect the particular position of mathematics as a gateway subject. For example, adults who wish to study science, technology, education and many social sciences in higher education must gain a recognised qualification in mathematics or its equivalent as an essential prerequisite for progression. Benn and Burton (1993) found that those people with a strong enough commitment and motivation to enrol on an Access course were not deterred by the mathematics on the course despite earlier bad experiences in learning this subject.

This analysis of who participates, though short, is crucial for our discussions around adults learning mathematics. It may be that the work-related nature of the discipline encourages participation by a wider group but it would be foolish to assume that the factors affecting general participation are not also forces on mathematics education. In addition, factors affecting participation in mathematics education may include the social context of mathematics itself, the potentially alienating nature of the discourse of the subject, where it is most comfortably and easily learnt, whether factors such as gender, race and class are central and if mathematics learning can help widen and increase participation in our society.

Adult education as a professional activity

Individuals who teach adults mathematics may or may not see themselves as adult educators and, though likely to be well educated, are less likely to have a formal qualification in adult education. This may be due to lack of opportunity or because either they or the system or both have not seen this as a valuable attribute. Any such qualification obtained is likely to be linked to teaching skills rather than critical enquiry into educational issues. The reflective practitioner approach is arguably a better basis for allowing the adult educator to acquire a more dynamic range of skills and knowledge. This is important because without some kind of critical reflection on professional practice, the educator becomes vulnerable to pressing outside forces.

The concept or occupation of adult educator is not easy to define due to the diffuse boundaries of adult education. Even the terminology is contested with some preferring terms such as tutor, facilitator and trainer. Houle (1960) made a useful three-fold division as: those who teach adults on a voluntary basis; those who perform this and other educational roles on a part-time basis for remuneration; and full-time adult educators. Many adult educators, whether full- or part-time, fulfil both teaching and administrative/organisational roles. Whichever way the role is viewed, the majority of people who are classified as adult educators undertake this role on a part-time basis (Jarvis 1983). A further twist is the perception of the individual as to her/his occupation. Some who teach adults will not regard themselves as an adult educator but as, say, an engineer, university teacher, or mathematician. Some will teach adults as just a part of their teaching commitments, others will be employed full-time in other occupations by which they will define themselves. Still others will only teach adults part-time but see this as their main role.

As far as the background and training of adult educators is concerned, there is a scarcity of systematic evidence. What there is shows that they are well qualified in their subject area but not in teaching skills (Lesne 1985). There have been influential people in adult education who have queried the value of education for adult educators. Houle (1960) asked whether those who were recruited for graduate training in adult education could measure up to those who have been chosen by the crude but effective self-selection which was then the rule. Whatever the answer, the positions in adult education are still filled primarily by persons with no academic training in adult education, who come into adult education by way of the proverbial 'backdoor'.

There has never been any requirement in Britain for adult educators to receive formal training in teaching skills. People are appointed on their qualification in their subject area perhaps on the assumption that knowledge of a subject ensures the ability to teach it. This reflects the strong assumption in our society that good teachers are born not made. Adult educators have therefore traditionally 'got by', learning on the job and improvising. Consequently many adult educators have had no or little formal education to encourage them to consider the hows of adult education let alone the whys and whos. In a sector experiencing so much change and turmoil, there is little opportunity or encouragement for critical consideration of educational purposes or the place of adult education within a wider social framework. This is perhaps particularly true for those whose own higher education was in disciplines such as mathematics where it is likely that their own learning experiences will have been firmly located in the positivist tradition.

Education of adult educators

The situation is changing with an increasing provision and take-up for the City and Guilds CGLI 7307 Stages 1 and 2 accompanied by a similar but smaller increase in Stage 3 provision. There is also a growth in Certificate of Education (FE) courses with remission for Stage 1 and Stage 2. These programmes together with other CGLI awards such as the Initial Certificates in Teaching Communication Skills (Numeracy) are the basis for a coherent structure of awards for adult educators (Foden 1992). In further education, colleges are increasingly demanding one of these teaching qualifications on appointment. However, these initial qualifications are increasingly competency-based and tend to focus on teaching skills. There is little time or space for consideration of ethical, practical and philosophical issues arising from the educational process. Criticising the new competency-based qualifications which are likely to dominate in the future, it is possible to search in vain for the advantages of this new type of training for the lecturers but see the result as the efficient production of willing, competent and compliant employees for the colleges. Lecturers trained in this way will be 'not only less likely to balk at what they are asked to do to their students but also more likely to display compliant, organisationally-appropriate attitudes and behaviours themselves' (Malcolm 1995). This worrying scenario reaffirms the urgent need for ways to promote engagement, understanding and creativity as well as competence amongst practitioners.

There are a growing number of Masters Courses in Continuing Education with a very few having options in Literacy and Numeracy.

Influences on adult educators: the theory

There is considerable debate as to whether adult educators have succeeded in their historical mission to promote greater social justice. Michael Newman recently examined the attempts on the part of adult educators to engage in radical activity (1994). He suggests that despite the claims of many adult educators, this has been very peripheral and halting activity. This is not only because, in some ways, it is very difficult work. It is also because of the total lack of support of the parent institutions. Indeed, especially at the present time, any attempt to deliver a coherent system of education for adults is marked for doom (Thomas 1995).

However, there may be other reasons. In order to understand the lack of success of adult educators, it might be helpful to recognise that adult educators are a sub-culture of society and to understand what Gramsci calls the 'popular consciousness' and the impact of the dominant culture (Mayo 1994). The 'spontaneous philosophies' of a group draw upon many influences and fragments of many ideologies. For adult educators, this may consist of the content of their degrees, their personal political ideologies, books which lead their teaching practice, the views of their colleagues, the ideologies prevalent within the institution and so on. These philosophies may find themselves at odds with the dominant culture. Fragments of ideologies or ideas may be absorbed by the educator to create a mental 'mosaic' which may be unsystematic, incoherent and subject to influence. It may even be contradictory (Benn and Burton 1995). It is these contradictory components which Gramsci suggests reinforce and shape the popular consciousness in a way in which allows groups such as adult educators to be seduced by the dominant educational cultures to greater or lesser degrees (Gramsci 1971).

Individuals interpret events through experience, social interaction and shared understanding, and it is this circle that accounts for structure and meaning. The creation of reality or Gramsci's mosaic can be seen as taking place through the process of building of mental maps (Rogers 1993). Each person through the interactive process creates knowledge for themselves. In conceptualising reality, practical activity and reality constructs can be seen in the same way as a navigational chart. This analogy shows that just as a navigational chart is always open to revision – so too are change, revision and 'openness' factors in the way in which reality and thus practice is constructed. The forces which cause the sands to shift include the major concerns of the individual: those things that are close to the centre of his/her map. This is a fluid zone and can change hour by hour as immediate concerns change. Things get pushed further away as other issues take a primary position. The map gets continually redrawn as issues become more or less value-laden. There are, in addition, longer-term changes as perception and understanding change. Things that were once at the outer reaches of the map move inwards towards the centre while other matters once central to a person recede and fade. The wider the concept distance between a force and the self, the less the influence.

Influences on adult educators: the practice

Work carried out by Fieldhouse (1993a) illuminates the changes over time in the mosaics of meaning of one sub-group of adult educators – university and WEA adult educators who came into the profession during the 1960s. The research builds up an excellent picture of the forces operating on adult education in the 1960s and how these had changed by the 1990s. Fieldhouse's work gives an insight into how adult educators order their mosaics of meaning, which are then modified by their differing experiences in education and pressure from dominant ideologies.

The 1960s were a time of protest and radical politics, illustrated by CND, the Anti-Apartheid Movement, the election of a Labour Government and student uprisings in France. The 'progressive' sixties ideology manifested itself in an optimistic expectation that the world could be changed for the better. Education was seen to play an important role in changing social conditions (Robbins 1963; Newsome 1963; Plowden 1967; Russell 1973). Fieldhouse found that this complex web of optimism, change and protest created a left-wing radical culture which was imbibed by those who entered adult education as tutors or organisers in the late sixties and early seventies.

But now different societal forces are operating and new reality maps/mosaics have been constructed. Fieldhouse's cohort has swung from centre-left to centre-right and earlier optimism about the efficacy of adult education for social change has somewhat evaporated. The move to the right in what were once idealistic tutors is a reflection of the widespread pessimism of the changed political climate in the eighties and nineties – the general depoliticisation of life and the growing selfishness and competitiveness at the expense of a belief in a general, collective good (Fieldhouse 1993a: 50).

The mosaics of adult educators are now heavily influenced by the current dominant beliefs that the role of education is for training and to reinforce the beliefs, assumptions and cultural traditions of the dominant ideology. This is reflected throughout the educational system and certainly throughout adult education, where recent changes in funding and in accountability reflect a major shift towards market-driven priorities (Benn and Fieldhouse 1993).

The vision of education as a tool to be used in the process of personal and collective consciousness-raising, the development of critical awareness and understanding, and the generation of 'really useful knowledge' and appropriate strategies for encouraging activism, responsibility and participation in society is hardly apparent today. The educational emphasis has changed to one of individual development (skills, competencies, goals), self-help and student-centeredness, with a narrower definition of development (Fieldhouse 1993b: 243).

These then are some of the forces that act on all adult educators in Britain today, including those who help adults learn mathematics. This discussion helps to clarify the forces acting on all of us to ensure that, without quite realising why, we may lose control over both our principles and practices. This may result in our seemingly willing contribution to an education system which has visions far different to our own.

The reflective practitioner

To counteract the mechanistic competency-based approach to the education of adult educators and the pushing-out of basic beliefs by stronger current ideological forces, the concept of the reflective practitioner as proposed by Schon (1982, 1987) is worth examining. Schon argues that being a professional involves not only the application of knowledge to instrumental decisions but also the use of a more dynamic form of knowledge. Here the professional is able to think constructively in context and reflect in real-time, not after the event, on the understandings which have been implicit in their actions, understandings which they bring to the surface, criticise, restructure and embody in further action.

It is assumed that the skilled professional has a repertoire of strategies and techniques, the capacity to think creatively in context and the abilities to apply formal knowledge. These facilities are all available to the practitioner at all times and are used at will to provide intelligent action. Hence the conclusion that professional activity is more than the application of knowledge and more than technique. It is an extremely complex mix of thought and action, of knowing and doing. So, any continuing professional development must be more than just the acquisition of further knowledge, important though this is. It should also provide the participant with the opportunity to develop further more complex skills. Critical reflection is fundamentally critical self-reflection, allowing the individual to step back from their involvement, view this in the context of a wider framework, and evaluate and consider alternatives. Through this kind of open critique, the individual reaches a position of intellectual independence in which they are enabled to see through the apparent inevitability of a proposition and its theoretical anchoring and so gain a measure of freedom from entrapment by any conceptual schema. It allows the professional to identify operative paradigms and consciously move between them (Barnett 1990).

The need for relevant and critical provision for the education of the adult educator

The ideas and practices of the reflective practitioner can be developed and incorporated into practice by the adult educator acting alone. However, as Chapter 1 argued, a programme of self-directed independent study, while meeting the individual's perceived needs, may preclude the consideration of critical social, cultural and political influences and therefore excludes social benefit and the collective good from any assessment of value. This mode of study may lack objectivity, allowing the adult educator's practice to remain subjective and normative. In the absence of a sustained social critique, the adult educator may become vulnerable to dominant pressures or ideologies. These are the very forces that the reflective practitioner approach is being employed to resist.

Space, time and opportunity are required to acquire this dynamic form of knowledge through group and probably formal learning opportunities. Formal training experiences for adult educators need to be widely available, more than competency-based, and could, with value, allow for the critical consideration of educational issues through approaches such as that of the reflective practitioner. All involved in the education of adults, whether termed adult educator or not, have much to gain by developing their professional skills through such training. Without this the critical appreciation of the wider context within which adult learn may be ignored or suppressed.

Section 2

The framework within which adults learn mathematics

Chapter Three

Mathematics: a peek into the mind of God?

Before further discussion of adults learning mathematics, we shall consider the nature of mathematics itself and also the place of mathematics within the broader context of human thought, experience and history. The fundamental tenet outlined in this chapter is that mathematical knowledge is a social construct and that mathematics is not 'a peek into the mind of God' but is created by human beings whose thinking is influenced by a historical and political context. This view of mathematics has profound implications for the teaching of the subject and places responsibility for the negative position of this subject in most people's minds on society, the subject and how it is taught rather than on the shortcomings of the individual.

First this chapter will examine whether mathematics is either rational thought evolving according to its own inner logic or an irreducibly social and cultural phenomenon, open to rational and irrational influences.

A time of certainty

For over 2,000 years, mathematics has been dominated by the belief that it is a body of infallible and objective truth, far removed from the affairs and values of humanity (Ernest 1991: xi, Benn and Burton 1995). This body of truth is seen as existing in its own right independently of whether anyone believes or even knows about it. Bloor (1973: 43) argues that this belief in the independent existence of mathematical truth implies that mathematics is a realm, a bounded territory. Knowledge and the use of mathematics then requires two stages, access to the realm and then activity within it. It is generally accepted that the first stage is fallible. Hence it is possible to discuss the process of selection and education and the influences which promote or inhibit access to mathematical skills. However, what happens within mathematics itself is regarded as closed to discussion. This is seen as predetermined and certain. Therefore a mathematical calculation is the tracing out of what is already there, the calculation exists 'in advance'. It was this belief in the certainty of mathematics which allowed Kant (1783) to write:

> We can say with confidence that certain pure *a priori* synthetical cognitions, pure mathematics and pure physics, are actual and given; for both contain propositions which are thoroughly

> recognised as absolutely certain . . . and yet as independent of experience.

The three categories that encapsulate the predominant modes of cognitive culture in the Western world can be seen as premodernism, modernism and postmodernism. Any understanding of the differing beliefs in the nature of mathematics needs to be seen in the context of these world views.

Premodern thinking was dominated by the belief in the authority of the church, the infallibility of the Pope and, through him, the total authority of the Church hierarchy and the Divine Right of Kings. This was reflected in society by authoritarianism, absolutism and certainty. Scholastic dogma of the time argued that knowledge was nothing but a mirror of reality. Primacy of geometric mathematics reflected this belief. This can be summed up by the following quote. 'Why waste words? Geometry existed before the creation, is co-external with the mind of God, is God himself' (Kepler as quoted in Koestler 1959: 264). These views dominated until the Reformation started a period of questioning that culminated in the Enlightenment.

Modernity developed out of the Enlightenment as a one-sided emphasis on formal rational instrumental orientations, overall disenchantment, the growth of administrative and technical rationality, the increased differentiation and specialisations of science, morality and art and a heightened sense of individuality. Modernism extols rationalisation, the objectification of knowledge and individual self-determination as part of the overall process of progress and enlightenment (Ashley and Betebenner 1993). Life is seen as the unfolding of progress through reason. This implies predictability, the generalisation of laws, and the universalisation of knowledge. Modernism is founded on notions of developmentism and progress where developmentism presents change as natural, regular and linear.

With change constructed as a universal, a powerful norm is created with the consequences that alternatives to this norm are closed off as pathological, marginal and invisible. Hence the story of human development becomes the 'grand narrative' of the dominant group in society and the narratives of other groups are ignored. The concept of progress is central to modernist thinking and is seen as a neutral description of reality and as giving both individuals and society a reassuring sense of meaning and coherence. In this process, with its associated sense of direction, difference is repressed in a demand for certainty (Usher 1995).

Mathematics has many, if not all, the prerequisites for modernity and has been rewarded for its rationality, generalisability, predictability, objectification and universality of knowledge by a central and powerful position in modern society. As a discipline, mathematics in the time of modernity seemed to mask reality by increased abstraction from the real world of human experience and endeavour. The modernist moved from God and geometry to mathematics as number. This was acknowledged amusingly in the quote 'God does not geometrize. He ever arithmatizes (*sic*)' (Jacobi quoted in Ashley and Betebenner 1993).

The relationship between mathematics and reality became increasingly problematic and puzzling, for example, the Banach-Tarski paradox (Ashley and Betebenner 1993). Some mathematicians were uneasy about this abstraction for its own sake. Others coped by adopting a mystic view that sought to grant some reality and objectivity to mathematics. But attempts to either picture the world as it is or use a perfectly consistent neutral meta-language ultimately failed. Indeed, in late modernity, mathematics appears to deliberately deceive by masking even awareness of the absence of any reality

Within the wider context of human thought and experience, the development of mathematics can be seen as the 'grand narrative' of academic Western mathematics which pathologises inability to relate to this mathematics and ignores or marginalises alternative or 'other' mathematics. Difference is repressed, the central narrative is held as certain and the workings of power are concealed. This characterisation of mathematics has provided an elaborate rationale and legitimisation for the pre-eminence of academic Western mathematics and has contributed to the dominance of certain cultural groups in society. The mathematical narratives of subordinate groups have been denigrated or ignored.

A time of uncertainty

To question the certainty of mathematics is to challenge the hegemony, irreversibility and sweeping narrative of modernity. It is to challenge the very foundations of modernism and hence only likely as the consequence of a major change in the perception of the nature of the human condition. One such change is the concept of postmodernity. Whether there is continuity or a radical rupture between the modern and the postmodern, current thinking postulates a new paradigm which takes into account the profound changes in economic, social and political life but also in different ways of knowing and understanding.

Writers such as Foucault (1973) and Lyotard (1994) see in the postmodern world the end of the 'grand narratives' or political theories, thus liberating space for a great diversity of narratives. Westwood (1992) highlights three crucial elements of postmodernism. First is European culture's awareness that it is no longer the unquestioned and dominant centre of the world. Second is the deconstruction of Europe with its totalising and universalising view and dominance. Third is Foucault's account of power as crucial to the construction of reality, language, meanings and truth, allowing for a multiplicity of discourses (Foucault 1977).

Within postmodern politics, categories are not static but are formed by power/knowledge relationships. Consequently, postmodernity by its emphasis on fragmentation and alternative world views, truths, realities and cultures rejects the realist and instrumental epistemologies and certainties of modernism. A postmodernist view of mathematics is based on the legitimacy of alternative mathematics rather than the dominance of Western academic mathematics; the belief in alternative truths and realities in mathematics and the construction of mathematics rather than its 'giveness'.

A critique of certainty

In the postmodern world, there have been serious critiques of the belief in the certainty of mathematics, the belief that fundamentally mathematics exists apart from the human beings that do mathematics and that 'Pi is in the sky'. Traditionally, the absolutist view of mathematical knowledge is that it consists of certain and unchallengable truths. But Lakatos (1976) shows that the quest for certainty in mathematics leads inevitably to a vicious circle. Any mathematical system depends on a set of assumptions, and trying to establish their certainty by proving them, leads to an infinite regression. Without proof, the assumptions remain fallible beliefs and certainty is lost. Axioms have to be accepted without proof, therefore certainty is gone. This leads to the view that mathematical truth is fallible and corrigible and hence open to revision.

However, the rejection of absolutism, though leading to a loss of certainty, is not a loss of knowledge, just knowledge in a different form. Bishop (1991) argues that mathematics has a cultural history. He cites the fact that triangles all over the world have an angle sum of 180 degrees but questions the 'universal' figure of 180. Why not 150, for example? The explanation lies in culture. The importance of the number 10 is surely a reflection of human physiology rather than a mathematical truth. Mathematical ideas have been decontextualised and abstracted and hence appear to be universal. However, it is arguable that mathematics is, in fact, a form of cultural knowledge which all cultures generate in the same way that language or religion are generated. This is supported by Kuhn's influential view (1970) that mathematics and science develop through revolutions in which dominant paradigms are overthrown and are about struggles around power and knowledge.

Mathematics as a social construct

So the certainty of mathematics has been under question. A growing number of mathematicians and philosophers are arguing that mathematics is fallible, changing and the product of human inventiveness (Ernest 1991). Others (Bloor 1973; Wittgenstein 1956) argue that absolute truth is located in utility and the enduring character of social practice rather than a calculation.

> And of course there is such a thing as right and wrong . . . but what is the reality that 'right' accords with here? Presumably a convention, or a use, and perhaps our practical requirements.
>
> (Wittgenstein 1956)

Wittgenstein grappled long with this problem but eventually placed the ontological claims of logic and mathematics in the same bracket as any other social institution. He concluded that the logic that we feel when we are performing the most inexorable of arithmetical practices is an acknowledgement of the repetitiveness of our habits or forms of life. We learn to count with endless practice, with merciless exactitude. That is why it is inexorably insisted that we shall say two after one, three after two, and so on. The truth in counting is hence pragmatic. The truth is that counting has proved to pay (Wittgenstein 1956: 4e).

Sociologists and mathematicians such as Ashley and Betebenner (1993) argue that philosophers have tried but failed to show how modern mathematics and science either pictured the world as it was or used a perfectly consistent, neutral meta-language. Mathematics did not develop in a cultural or social vacuum but rather it reflects and magnifies cultural transformations. Mathematics is not a body of truth existing outside human experience. It is a construct or an invention rather than a discovery; a collection of norms and hence social in nature. Hersh (1986: 25) regrets the loss of certainty but still argues against the attempt to root mathematics in some non-human reality and for the acceptance of the nature of mathematics as a certain kind of human mental activity. The result would be a loss of some age-old hopes but a clearer understanding of what we are doing and why.

Implications of the move from an absolutist to fallibilist view

Ernest (1991) identifies three main differences between these views. Firstly, absolutist thought sees knowledge as a finished product but fallibilist thought sees it as a continual process of generation and renewal. This implies mathematical knowledge lies within the human context of knowledge and its historical genesis. Secondly, absolutism sees mathematics as the only realm of certain knowledge which leads to the view of a separate and isolated discipline whereas the fallibilist view is that mathematics is embedded in human history and practice and is hence an integral part of the body of human knowledge. Lastly, the absolutist views mathematics as neutral and value-free as opposed to subjective and value-laden.

This thinking led Lakatos (1976) to develop a philosophy of mathematics founded on the belief that mathematics is what mathematicians do and have done, with all the imperfections inherent in any human activity or creation. Mathematics is founded on mathematical practice and is a dialogue between people tackling mathematical problems. As a human activity, it is subject to change and error and it must be viewed within the wider context of all human activity. Mathematical activity produces mathematics. This approach views mathematics as a social construction for the following reasons. Mathematical knowledge is based on linguistic knowledge, conventions and rules, and language is a social construction. Inter-personal social processes are required to turn an individual's subjective knowledge into accepted objective mathematical knowledge, and objectivity itself is social. This leads to the opinion that mathematics changes over time and is an agreed rather than objective body of knowledge.

Cobb (1986) notes that self-generated mathematics is essentially individualistic so it is in a sense anarchic mathematics. In contrast, academic mathematics embodies solutions to problems that arose in the history of our culture. Unless individuals intuitively realise that the standard formalisms are an agreed upon means of expressing and communicating mathematical thought, they can only assume that these are constructed as arbitrary dictates of an authority. Academic mathematics is hence totalitarian mathematics.

Those with experiences of teaching adults may recognise the residual belief in absolutism in many of their students through their total alienation from ownership of mathematics and the 'mysticism' that the unrooted nature of the subject engenders. An absolutist approach to mathematics sees the subject as owned by others. The fallibilist approach sees it as belonging to everyone, and all are engaged in the learning and doing of mathematics. By transferring ownership, more than just mathematics is involved. If mathematics is neutral and objective, it disempowers individuals from mathematics and for the social, political and economic uses of the subject and hence control in this society. Ownership of mathematics allows critical awareness of society and hence increased power.

Mathematics as neutral and value-free

The attack on the certainty of mathematics led to the questioning of its neutrality. If mathematics is certain, if it reflects the God-like power of innate, transcendent human reason, if it is a body of absolute truth, and if the answers are already written, then it is independent. It must be neutral. However, if mathematics is a social construct, an invention not a discovery, then it carries a social responsibility. Absolutist philosophies are committed to the objectivity and neutrality of mathematics. However, mathematics does have implicit values – the abstract is valued over the concrete; the formal over the informal; the objective over the subjective; justification over discovery; rationality over intuition; reason over emotion; the general over the particular; theory over practice; the work of the brain over the work of the hand (Ernest 1991).

Though this accusation is refuted by absolutists arguing that these values concern mathematicians but not the realm of mathematics itself, once these rules are established they impart values to the subject. The fallibilist, however sees mathematics as the outcome of human activity and hence culture-bound, value-laden and embedded in its cultural context.

Culture free/culture bound

A proponent of this view, Joseph (1987: 22–23) suggests that the present structure of mathematics education is Eurocentric and modernist, being based upon four histographic pillars:

- the general disinclination to locate mathematics in a materialistic base and thus to link its development with economic, political, and cultural changes;
- the confinement of mathematical pursuits to an elite few who are believed to possess the requisite qualities or gifts denied the vast majority of humanity;
- the widespread acceptance of the view that mathematical discovery can only follow from a rigorous application of a form of deductive axiomatic logic believed to be a unique product of Greek mathematics; hence, intuitive or empirical methods are dismissed as having little mathematical relevance;
- the belief that the presentation of mathematical results must conform to the formal and didactic style devised by the Greeks over 2,000 years ago and that, as a corollary, the validations of new additions to mathematical knowledge can only be taken by a small, self-selecting coterie whose control over the acquisition and dissemination of such knowledge has a highly Eurocentric character.

Many writers (Joseph 1987; Anderson 1990; Bishop 1990) argue that the Eurocentric bias of mathematics infuses the subject with an elitist, racist and sexist bias. They argue that the belief in the certainty and neutrality of mathematics and science deprives these subjects of any cultural or social context. Hence mathematics and the natural sciences place no value upon the historical, cultural or political milieu within which they are located. Indeed mathematicians such as Pythagoras, Euclid, Cauchy-Rieman, Fourier, and Newton are cited as the source of Western mathematics without any further reference to the times within which they lived or to the influences upon their work. They are abstracted from time and space and presented as if they and their work are timeless, complete and absolute. This separation from culture and relevance makes mathematics inaccessible to those already alienated from society by educational disadvantage and by class, race and gender.

So we have outlined two incompatible views of mathematics. One is premised on certainty, neutrality, the peek into the mind of God. The other sees mathematics as a social construct and hence open to change, progress and development and as an unfinished project. These differing views lead to a fundamentally different approach to mathematics teaching and learning and hence different attitudes of adults to learning mathematics.

Reluctance to accept the fallibilist view of mathematics

As argued by Davis (1986: 64), the reception given to proponents of fallibilism still ranges from coolness to indifference. The absolutist belief is deep in the psyche of mathematicians, learners and teachers and its influence still distorts mathematics education. There are several value components of culture which affect attitudes to mathematics and tend to preserve the absolutist view of mathematics (Bishop 1991). Humanity tends to value security, and the power of mathematics in our society derives partially from the feelings of control, security and certainty that mathematics with its 'right' answers gives.

The privileged position of mathematics or numeracy in the 'back to basics' campaign is arguably less to do with mathematics *per se* than with its aura of certainty. Similarly, mathematics gives people a sense of progress. A mathematics problem is solved, the process abstracted and generalised, and hence other problems are solvable. The perception is that knowledge has been developed. Absolutist mathematics aligns with many fundamental beliefs and philosophies through its rationalism, logic, objectivism and reasoning which are all valued so highly in modern Western culture. This guarantees the significance of mathematics in our society (Walkerdine 1988).

A further strength of mathematics lies in its seeming openness to examination by anyone whilst it also retains a sense of mystery. However, the openness of mathematics is illusory. Notions of common sense and rationality are expressed by the dominant culture through culturally-specific discourses. The dominant rationality is that education is based on narrowly-defined boundaries for knowledge and reason measured against particular norms, and ignores the logic expressed through alternative discourses. Adding a political component whereby mathematics is seen as an 'invisible' force operating to help preserve the *status quo* and perpetuate divisions in society, it becomes clear that, for many, the absolutist view of mathematics is safer and more comfortable than that of the fallibilist.

Failure of absolutism in mathematics education

The British education system still fails to provide a substantial proportion of the population with even basic mathematical skills (see Chapter 8). Even worse, it leaves many with an abiding dislike of the subject. Mathematics education will now be examined in the context of the belief in the certainty and neutrality of mathematics for part of the explanation of this failure.

The concept of mathematics as a body of infallible and objective truth, whilst questioned by many mathematicians and philosophers, appears to be still widely held by society, teachers and students. An analysis of both the Cockcroft Report (1982) and a report by Her Majesty's Inspectorate which looked into the nature of mathematics teaching in Britain (1985) concludes that an absolutist view of mathematics is assumed (Ernest 1991: 223).

This clearly has a number of consequences for the teaching of the subject. Abstracted from any socio-political context, mathematics can be taught within the strictures of its own boundaries thus retaining for the learners its mysticism and ritualistic nature. Certainly much work has been done its introduction into the mass education system to increase understanding of mathematics. However, just as certainly, the history of mathematics is one of failure on a large scale for its students. The Cockcroft Report notes that at the time the report was written 'about . . . one-third of the year group, leave school without any mathematical qualifications at O Level or CSE' (1982: 56) and Chapter 8 shows that this poor level of performance is still with us.

It seems that despite calls for over 100 years for an approach to mathematics that interests and stimulates children at school, mathematics is still a subject that confuses, alienates and leads to failure. A school inspector wrote as recently as 1990 that she was horrified to find that at both primary and secondary level 'nobody seemed to enjoy mathematics; not even the teachers' (Cross 1990: 4).

Writers such as Rogers (1969), Dewey (1964) and Knowles (1980) argue that learners are self-directed beings who learn best when they perceive the relevance of knowledge to their lives, and when learning is related to problem-solving. If mathematics is perceived as a fixed and unvarying body of truth independent of social concerns, then it is difficult to see any room for negotiation or where life experiences can be used in the learning process. If mathematics is neutral it has little to contribute to the learner's knowledge of themselves or their immediate world. All this contributes to a lack of motivation and hence a tendency to failure.

Mathematics education, culture and values

Historically society's view of mathematics has been grounded in logical positivism: mathematics consists of formal systems and is not verifiable by reference to experience. The body of formal knowledge does not change or grow. Hence this mathematics is removed from human activity and everyday experience. This, together with abstraction, makes it more and more inaccessible and sustains the elitism of the subject. Over recent years this has changed and now mathematics pedagogy is grounded in activity, discussion, investigation, questioning and hypothesis.

However, the culture is still dominated by the cult of individualism as opposed to collectivism with its valuing of shared experience, co-operative learning and the social construction of knowledge. Mathematics in formal curricula is still often perceived as preconceived structures to be developed, not knowledge to be constructed. The National Curriculum, with its emphasis on assessment, limits teaching styles and approaches and appears to be causing mathematics teaching to revert to a curriculum that is hierarchical, abstract and removed from reality (Ernest 1991).

There are strong cultural forces at work that impose a view of mathematics on all of us. Our schooling has encouraged us to accept as unproblematic that traditional mathematics somehow embodies uniquely powerful knowledge and eternal truths. This body of knowledge is not only infallible but also universal. It is well-defined, culturally neutral and value-free.

Constructivists, however, view mathematics as a human creation evolving within human contexts. In this view, the mathematics of cultural groups, such as national tribal societies, labour groups, children of a certain age bracket, professional classes and so on (which D'Ambrosio has usefully termed ethnomathematics) is as valid as academic mathematics (D'Ambrosio 1991).

Academic mathematics can be seen as just one ethnomathematics, albeit a powerful one.

In the postmodern worldview, cultural and structural alienation from mathematics can be acknowledged, as can the social results of success (or otherwise) in mathematical endeavours. Society can recognise the diverse cultural and historical origins and purposes of mathematics and the contribution of all cultures to these. This approach would legitimise ethnomathematics and identify the history of mathematics as a record of humanity's struggle to problematise situations and solve them by constructing mathematics. If enculturalisation is the induction into the individual's own culture and acculturalisation is the induction into a culture which is in some sense alien, then, for a numerate society let alone social justice, mathematics teaching needs to be as much as possible the former not the latter. Many, if not most, learners are confronted with the mathematics of a subculture which they are not and have no wish to be members and where there is no cultural resonance (Mellin-Olsen 1987).

Implications for mathematics education

Constructivism is an approach rather than a way to teach. It is based on the premise that each person constructs their own knowledge where the raw materials come from the individual's prior experiences. The mental building blocks are made up of constituent parts that are for each learner both personal and idiosyncratic. However, social interactions are crucial in the construction of knowledge. Groups share meaning and the social perspectives of common lifestyles. In what way, then, do learners 'come to know' and how can educators help in the process?

Within the constructivist paradigm, the role of the tutor can be seen as providing experience. They can assess and estimate the knowledge that the learner has built by observing and listening, giving opportunities for organising, talking about and representing ideas and encouraging mapping of representation to develop modes of enquiry. The tutor can keep discussion open, revisit ideas and seek opportunities for generalisation and extension. Learners can be encouraged to produce co-operative ideas and support all ideas with suitable justification and argument (Maher 1996).

This is not an easy approach. It raises fundamental questions about knowledge, teaching and learning. Questions such as should the tutor 'tell' the learner anything? It also creates tensions between for the tutor as provider of an environment with a wide range of opportunities but not in control of the outcome of the learning process. It raises questions as to what it means to 'know' something and how the tutor can tell if it is known. Many adults come back to education expecting to be told.

Clearly, the constructivist approach can lead to conflict between the tutor and the learner, and the tutor and the educational establishment. Realistically, the tutor must walk the fine line of managing the learning experience so that the learners' expectations, whatever they may be, are met whilst sensitively offering mathematical challenges aimed at the conceptualisation of mathematics. An example of using this approach in time measurement could be through examining how units of measurement evolved through physical and historical conditions (for example, rotation of the earth around the sun) and collecting a list of sayings about time and tracing their origins (for example, saving time, wasting time). This would show the subjective, socially-negotiated and problematic nature of the concept of time (NSDC 1995). This approach would inform the whole learning experience. For example, if a learner does not understand a concept in a positivist classroom, the tutor might ask whether s/he had not expressed her/himself clearly or whether the learner had not listened carefully. In a constructivist classroom, the tutor would consider how s/he could understand the learner's construction or understanding of the topic in question.

We have stressed the value-laden nature of mathematics and hence mathematics education and denounced the practice of concealing these values particularly in adult education. We have argued that little attention is paid at present to values and consequently values are learnt but learnt implicitly, covertly and without conscious choice. Education should make values explicit and overt in order to develop the learner's capacity for choosing.

One question that might be asked is whose needs are being served by the present system and whose disregarded. The probable answer is that those who historically have benefited are the white middle class males, those who are disadvantaged are women, lower socio-economic groups and some minority ethnic groups. If this is correct, then the social construct of mathematics can be seen as another hegemonic force to preserve the *status quo,* confirm existing privilege and exclude from positions of power groups that have traditionally been excluded. Mathematics education needs to explicitly acknowledge values associated with mathematics and its social uses. Values need to be defined, for example, to include social justice, equal opportunities and democracy. Learners need to be aware of implicit social messages in the curriculum and have the confidence and skill to understand the social uses of mathematics.

A set of educational principles also needs to be devised. These can be taken from the writings of theorists. For example, Ernest suggests that if mathematics is a social construct then learners should be empowered to create their own knowledge and the mathematics curriculum should be centrally concerned with human mathematical problem-generation and -solving and should reflect fallibility (1991: 165).

Alternatively, and more powerfully, groups of mathematics educators can devise such 'mission statements' for themselves. This is precisely what was done by the Massachusetts Adult Basic Education Team. They first wrote a mission statement and developed and made explicit a written belief system, and then devised *from* these a set of mathematics standards and implementation approaches (Leonelli and Schwendeman 1994). Others have adopted their work, but the group argue that the main benefit lies in the development process and that educators should be encouraged to set up their own groups.

The concept of mathematics educators collaborating in the process of agreeing or at least discussing a set of values and principles to underpin mathematics for mathematical education though heartening is problematic and raises difficult issues. Whose values and belief systems should be used, the group's or society's? Woodrow asserts that 'on many social issues, the teacher's own beliefs may need to be subsidiary to those of the society he/she is called upon to represent' (1989: 233). One of the real values of working with a group as described above is the opportunity to explore these moral and ethical issues in a safe and supportive environment.

To open up mathematics so that all can engage requires an active construction of understanding built on the learner's own knowledge and an exploration of the learner's own interests. The attempt to convey ideas and concepts to the learner must take place using the metaphors and imagery available to the learner. These are clearly the consequences of the society and culture within which the learner lives.

If mathematics is a dynamic, living and cultural product, the contextualisation of problems is essential. Just as Freire (1972) started with strong political words to engage his learners' passionate interest, so we need contexts which are genuinely meaningful to the adult learner. We need to use the language and culture of the learner so that the mathematics becomes their own. Freire's concept of education for liberation places knowledge and power in the hands of the learner, whereas the banking concept places control with the elite. It is clearly to the benefit of the elite to preserve the latter approach rather than the former which transfers control to all.

In mathematics education, the kind of questions asked and examples used all convey political, social and cultural attitudes. The fact that these are covert makes them all the more insidious. Attitudes to race, gender and class in mathematics can all affect individuals' perception of themselves. In this new fallible and constructed mathematics, contradictory ideas become competing theories as mathematics knowledge develops. Objectivity is arrived at by taking into account the accumulated experiences of all periods and cultures and hence reaching an objective point of view in terms of being neutral (Lerman 1992).

This fallibilist approach allows mathematics to be seen in terms of growth, change and context. It implies more than looking for aspects of different cultures to exemplify mathematical ideas and includes looking at how cultural perspectives may affect mathematical modes of thought. Re-examining beliefs about mathematics and considering it in terms of its social nature and foundation is very complex and problematic. If culture is defined as groups of individuals bound together in a social context, the problem may be more easily resolved as catering for cultural individuality in a shared setting. Bishop (1991) identifies six categories as the mathematics that all cultures do – counting, locating, measuring, designing, playing and explaining and suggests that mathematics as cultural knowledge derives from humans engaging in these activities in a sustained and conscious manner. The foundation for a new mathematics curriculum could be located in these categories.

This chapter has examined Western views of the nature of mathematics, locating these in the broader context of the history of Western thought. Whilst acknowledging the continuing power and influence of the absolutist view, it argued that the way towards a more egalitarian society lies in that of the social constructivists. The 'grand narrative' of Western academic mathematics needs to respect and allow space to alternative local narratives.

Chapter Four

A matrix of factors

The aim of all education, including mathematics education, is to enable learners to satisfy goals such as vocational and personal development but also facilitate and encourage learners to participate fully as citizens. In a democratic society, this implies curricula that serve everyone in that society, with aims and objectives located in human and social good and which are not just consumer-driven, corporate or reproductive. The thesis of this book is that by this criterion, mathematics education at all levels alienates and fails a large proportion of the population but that it is possible to start to change this situation by locating all mathematics education for adults in a philosophical, political, historical and social framework with a curriculum and pedagogy informed by this conceptualisation.

Learning mathematics

What learners learn is a result of attempting to solve their own problems: people learn mathematical knowledge if the problems that they have are mathematical. Adults will not learn mathematics if they see mathematics problems as someone else's problem, whether the someone else be the tutor or sections of society with which the learner does not identify. But adults, either individually or in groups, do have mathematical problems in their everyday life. This results in not just one mathematics but many.

As well as the mathematics that is commonly taught in educational institutions (academic mathematics), there are many other forms of mathematics that have been devised by different groups to meet the needs of their own cultures. Each has its own discourse and is valid and legitimate within the cultural group. These can be called local or ethnomathematics: the mathematics which are practised by identifiable cultural groups. Culture here is defined widely to include not only gender, class, ethnicity and age but also vocational groupings (D'Ambrosio 1991). The set of all local or ethnomathematics includes the set of formal academic mathematics which is the ethnomathematics devised and owned by the powerful and dominant in our society. This mathematics dominates to such an extent that in Western society it is seen as the only legitimate mathematics. Consequently there is an overwhelming tendency for the formal mathematics curriculum to be concerned only with academic mathematics and not local or ethnomathematics. Hence, the role of mathematics education is often seen as giving the learner fluency in some required aspects of formal mathematics by building on the learner's existing knowledge in academic mathematics only . This process is illustrated in Figure 1.

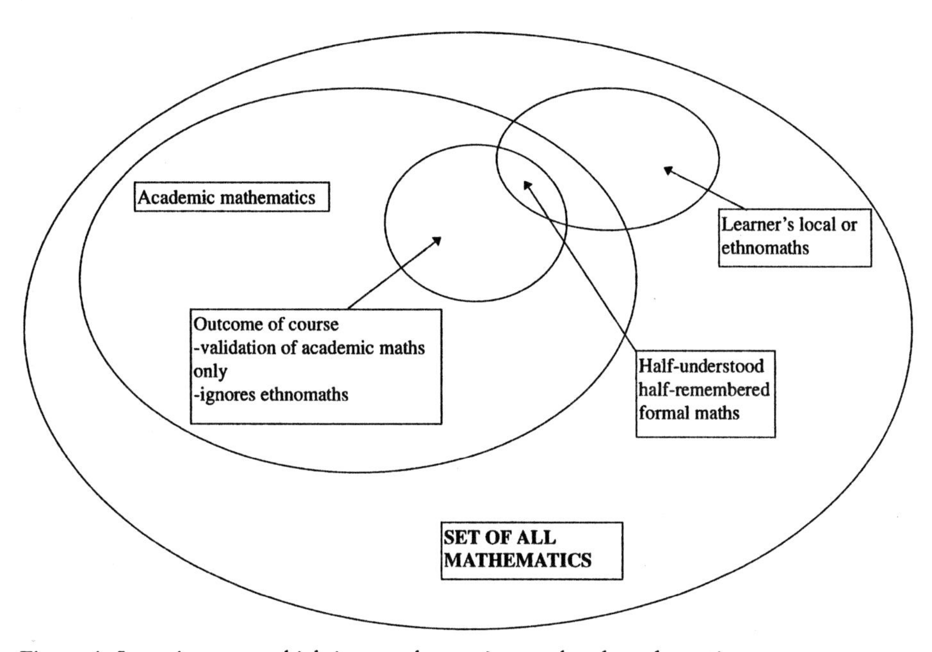

Figure 1: Learning map which ignores learner's own local mathematics.

Superficially, this makes pedagogical sense. After all, a comprehensive grasp of academic mathematics ensures that the codification from real world problems to mathematical abstraction and *vice versa* causes no serious problems. Local or ethnomathematics, being the mathematics used by only a certain section of society, are limited – hence codification both ways can cause serious problems. It might reasonably be concluded that it is 'better' pedagogical practice to continue to ignore ethnomathematics. However, as Chapter 8 will show, this approach has not worked, or rather has only worked for a segment of society. Many people never acquire the requisite comprehensive grasp. For some (white middle class males from professional backgrounds) there does appear to be a good correlation between their ethnomathematics and academic mathematics. Academic mathematics is, in reality, constructed on the needs and problems of this group and is therefore their own local or ethnomathematics.

However, for other more marginal groups such as women, minority ethnic groups and lower socio-economic groups the story is very different. In a worst case scenario, there may be little or no overlap between the syllabus of academic mathematics and the learner's ethnomathematics. The likely outcome is disenchantment and failure or, at best, the mathematics learnt is supported by a very shaky foundation. This approach pathologises the learner and designs techniques that will change the learner's behaviour and inculcate coping skills to make up for what are claimed to be objectively identified deficiencies (Collins 1991). This shows disrespect for the experiences and knowledge of the learner and disrespect for their potential power. The knowledge of the learner remains buried and invisible.

An approach based on Figure 2 may rectify this.

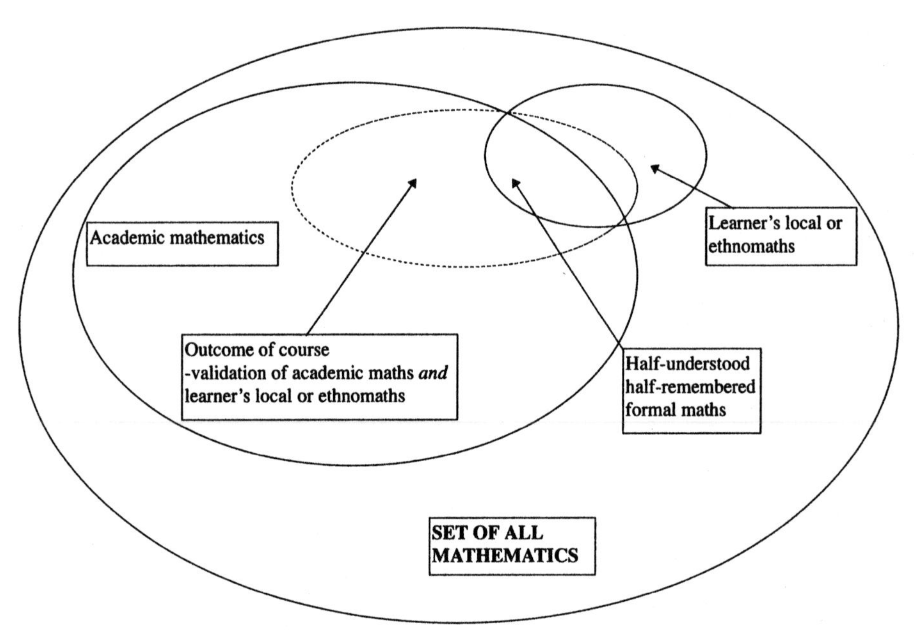

Figure 2: Learning map which builds on learner's own local mathematics.

Here the curriculum incorporates and builds on the firmer foundation of the learner's own ethnomathematics and, by use of imagery and metaphor, builds from what is understood and valued by the learner into what is not understood but is desired by the learner. With ownership of two mathematical discourses, the learner can move freely between the permeable boundaries with the ability to choose in a given circumstance which it is appropriate to use.

This approach respects the learner's knowledge and contributes to the development of both this knowledge and the learner's empowerment. It builds on the mathematics that the learner has previously acquired to solve their own problems. It more effectively allows the learner to achieve mathematical aims through the acquisition of more mathematical tools to solve the learner's mathematical problems. It also allows the learner to achieve social aims through the acquisition of qualifications in academic mathematics to solve the learner's social problems.

The learning continuum

But in a democracy, adults need to learn mathematics not only to develop skills to generate and solve their own mathematical problems, nor just to gain qualifications. They also need to understand why and how mathematics is generated, used and maintained in our society with concomitant consequences for democracy and citizenship. Every curriculum lies on a continuum as shown below, with banking education for control, reproduction and conformity to the *status quo* on one end and emancipatory education for democracy, independence and self-direction on the other.

Emancipatory education ← *The curriculum* → Banking education

Ultimately education can lead to reproductive or liberatory change: the former domesticates learners by simply helping them to adjust to socially-expected development tasks whilst the latter assists them to question fundamentally their perspectives on the world and their place in it (Tennant 1994). There is always a belief and value system at the centre of any curriculum. At present, the system is implicit and tacit. The dominant forces are located in a market-driven reproductive system whose prime purpose is the continuation of the *status quo* and the economic imperatives of Britain plc.

In a democratic society, value systems should always be explicit and mathematics educators need to move towards the construction of alternative value systems for the formal education of mathematics for adults. A more democratic humanist approach would require a critical evaluation of the socio-economic and political realities of society and endeavour to allow individuals more effectively to be in control of their environment and lives through their own understandings and actions. Whether adults seek to learn mathematics for utilitarian work or life reasons or to be more in control of their lives or to achieve their own social goals, they will be better able to achieve their ends with a clearer understanding of mathematics and their relationship to it.

The learning pathway

The prime purpose of the mathematics classroom is to allow the learner to achieve their stated objectives. The educational process is the transition from the learner's present knowledge of mathematics to the achievement of the learner's goals, whatever these might be. However, the process may well be ineffective unless and until the learner can locate themselves and their experiences in the wider framework that encompasses philosophical, historical, cultural, political and educational factors. Figure 3 illustrates the steps necessary to achieve effectively the required transition.

This is not to argue that raised awareness of these forces is sufficient to produce transformations of social reality. The critical reflective approach, which

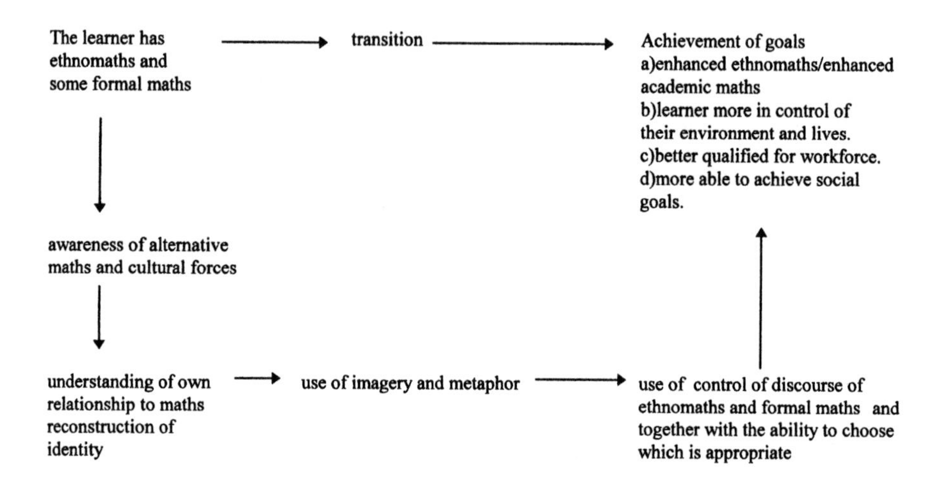

Figure 3: The learning pathway

this book argues should be an outcome of the curriculum, may be rejected by the learner, be accepted in the context of the classroom but not transferred elsewhere in the learner's life, or result in frustration and anger on the student's part over an increased awareness of forces and circumstances which the individual cannot change. An emancipatory intent is no guarantee of an emancipatory outcome. This is not to argue against this more emancipatory form of education: far from it, but educators need to be aware of these potential outcomes. A realistic assessment of the outcomes of any change of teaching is that individuals might be more frustrated in their new state of consciousness but they may also be more empowered to join attempts to change to a more just society. The net result in our consensual society will not be revolution but might lead to a slight shift in the political climate towards a more democratic and participatory society.

A matrix of factors

The powerful forces operating on the three main actors in the learning and teaching process – the learner, the tutor and the curriculum – can be represented by the matrix shown in Table 1.

All elements in the matrix are vectors *ie*, variable with both direction and magnitude. Each is acting on the learner, tutor or curriculum with a push or pull factor of varying strength towards either an emancipatory, empowering education or a banking, reproductive one. No variable is intrinsically more important than any other: each has an impact. The strength and direction will vary over individuals, institutions and societies and over time. The purpose of the book is to reconceptualises the process of adults learning maths in terms of this matrix. We will examine this framework from the position of the key actors – the learner, the tutor and the curriculum – then apply this to the development of the curriculum.

Table 1: *Factors acting on the learner, tutor and curriculum*

	Learner	Tutor	Curriculum
Goals	personal development; utilitarian; social purpose.	professional satisfaction; learner's goals; reconstruction of part of tutor's mathematical knowledge.	to enable learning; satisfy value system.
Cultural forces	attitudes to gender; class, race, age etc; current world view; horizons of time and place.	attitudes to gender, class, race, age, etc.; current world view; horizons of time and place.	attitudes to gender, class, race, age, etc.; current world view; horizons of time and place.
Political forces	political expectations; climate of lifelong learning; qualification 'mania'.	political imperatives of government, quangos, funding bodies and institutions.	political imperatives of government, quangos, funding bodies and institutions.
Educational forces	experiences of school and life.	experiences of school, education and life; dominant educational theories; institutional educational beliefs.	dominant educational theories.
Experiences or philosophies of mathematics	experiences of formal mathematics; ethnomathematics.	personal philosophy of mathematics.	philosophies of mathematics.
Learner	reflective learner.	strong motivation of adult learners.	articulation between learner's goals and curriculum.
Tutor	pedagogical approach.	reflective practitioner.	articulation of tutor's pedagogy and philosophy with that of the curriculum.
Curriculum	match of curriculum to learner's goals.	articulation of tutor's pedagogy and philosophy with that of the curriculum.	–

Chapter Five

Locating the learner

The learner and personal goals

Adults come to learn mathematics in formal education to satisfy a wide variety of goals. The triggering factor is something in their lives that they wish or need to do: they have a problem, either social or mathematical, which needs resolution but do not have the required mathematics. In community education some of the most common reasons for attendance include: to gain qualifications in mathematics (and English) as entry requirements for further study; some need extra mathematical knowledge to deal with a work problem; the self-employed often need skills for book-keeping, *etc*; younger adults sometimes wish to help their children with school work; older people sometimes need very basic mathematics to read recipes, understand decimalisation or cope with the 24-hour clock; men who have been made redundant sometimes realise that they do not have the basic skills to apply for other jobs.

A similar pattern of reasons for adults learning mathematics occur in further education. Here the mathematics provided may be at a higher level but still the triggering factors for the majority of adult learners is the need for mathematics, and often a mathematics qualification, for further education or vocational purposes. Some come of their own volition, some are sent by employers or educational guidance. Some again study with their children, particularly when courses are provided in the evenings or the weekends. Even adults who have entered higher education need help with their mathematics and higher education institutions often provide mathematics support workshops for this purpose.

Many also come 'for themselves' *ie*, for personal development. In our society, both the knowledge-that-is valued and qualifications-that-recognise-this are owned by others and are hence social rather than personal constructs. Nevertheless, it is arguably the personal development factor which gives 'real' meaning for the learner. Unhappily, the low levels of confidence and competence in mathematics in society generally has led many adults to 'deny' the importance of the subject in their lives and sometimes to redefine personal goals solely to continue to exclude mathematics from their lives.

The learner and political forces

There is a growing political awareness of the importance of a well-educated workforce. To encourage a flexible workforce which can adapt to the demands of a rapidly changing environment, successive governments have placed increasing emphasis on lifelong learning (DfEE, the Scottish Office and the Welsh Office 1995). This has resulted in secured funding for vocational courses for adults and a substantial expansion in both further and higher education. For demographic reasons, much of this expansion has been filled by adults. The financial emphasis on part-time students reinforces the bias towards adults. The funding methodology has affected fees. In many further education colleges, basic education courses are free whilst some are making courses free to anyone on state benefits or even free to everyone.

There has been a rapid increase in competency-based vocational qualifications, many of which are gained in the workplace. This has resulted in a growing expectation by adults that education is not just for the young. It has also led to an escalating demand from employers for higher and higher qualifications for all jobs. Mathematics is often used as a gateway subject, not because of the requirements of the job but to illustrate a level of cognitive skills. All these factors have led to a climate of lifelong learning and a qualification-hungry society which have affected adults learning at all levels. The political forces have led to a social climate where returning to study is seen by more adults as natural, almost inevitable.

The learner and cultural forces

Adults are creations of their own time and place. Within all societies, there are multiple cultures so adults are confronted with alternative and competing cultures and cultural traditions. A fundamental challenge for every adult is to learn to function within these possibly very different cultures. Culture is the medium which provides the important link between the individual, their learning and the social environment. People discover, create and use culture. They create their own meanings by themselves and with others. Hence learning is a social transaction as well as a solo performance. The transactional nature of learning means that the learner is usually inducted into a culture by more skilled members with adults coming to formal education partly because they feel that they will do better if guided by an expert. Any contemporary learning strategy has to incorporate within it the existence of different cultural groups and the notion of cultural difference and diversity. At the same time, the learner has to modify or adjust their individual culture to accommodate the demands of the new culture of the formal learning. The extent to which this can be done depends on the understanding by both tutor and learner of this dual system (Zeldin 1992).

However, all cultures are not valued equally. Our society has created inequalities between groups and the learning which adults gain through life is structured on inequality with learners positioned by society (Stuart 1995). This structural inequality will have influenced all adults and will have shaped their self-concept, their experiences of learning and learning mathematics in particular. They will bring the results of these experiences into the mathematics classroom.

The learner and experiences of education, mathematics and learning mathematics

There is a distinction between education and learning. Society, through its educational institutions, has control over the definition of education and its practice. However, there is a learning-in-practice which is not in the domain of the educational institution but is controlled by individuals as they go through life. Hence, knowledge acquisition and understanding cannot be limited to any one educational definition of learning. In addition, notions of education are not fixed: they are socially constructed within structures of power which exclude the majority of learning experiences.

Educational definitions focus on learning which occurs in institutions and within clearly defined subject areas and within specific curricula. This disregards the fact that by the time adults come to formal learning situations, they have accumulated a vast amount of knowledge outside institutions. However, those who missed out educationally the first time round or who have been badly failed by the system may place greater emphasis on formal learning and the skills and educational capital that others possess and which they feel they lack. Feelings of inadequacy and inferiority may be heightened on return to formal education as learners denigrate their own experience and insights. While adults with greater levels of formal education may take a more balanced view, almost all learners appear to regress and deprecate themselves for a while (West 1996; Weil 1993). Lecturers may disparage the personal as anecdotal and unreliable in comparison to more objective understanding and thus lose the rich interpretative possibilities of adults' experience.

The reasons for this are at least on part cultural. White male middle class values count as what is valid knowledge and acceptable discourse; reason and rationality are valued above feelings; and the more formal scientific way of knowing is valued against the emotional and the experiential. Personal and public worlds are separated and local ethnoknowledge is devalued (West 1996; Flax 1990; Seidler 1994). Present experiences are determined significantly by past experiences which have contributed to the way in which the learner perceives the world. Positive associations with learning will contribute to a positive involvement in new learning experiences. Negative experiences will work against present learning.

An illustration of the foundations on which many adults' mathematical knowledge is built can be seen from the results of research in the USA on understanding of percentages (Ginsburg and Gal 1996). To investigate what knowledge adults bring to the class, they started with such everyday use of percentages as '50% of' and '100%' apple juice. They found, for example, that it was not uncommon for people to assume that 50% of 250 is 125% and that 100% apple juice meant *mainly* apple juice. They concluded that people have a mathematical idea of 100% *and* a reality idea of 100% and, for many, 50% is a word, a colloquial understanding *not* a mathematical understanding. They concluded that people have a very patchy knowledge tied to experiences and as all adults have different experiences so they have different mathematical knowledge.

The implications of this are that people construct meaning within a personal framework. As a result, they can sometimes compute but do not understand, which leads to consequent problems of transfer. People also sometimes understand socially but not mathematically. Teaching adults mathematics in the classroom as if it is objective and context-free without unraveling these complexities of meaning or exposing the social structure of each individual's mathematics is unlikely to lead to anything but superficial success.

A useful tool for understanding how the experiences of life, education and learning mathematics have affected the learner is to use the concept of reality maps (Rogers 1993). This starts from the premise that all knowledge is the result of our own active mental process: it is a personal construct. Each person engages interactively with their physical and social environment creating knowledge for themselves. Learning is not finding out what other people know but solving our own problems for our own purposes by questioning, thinking and testing till the solution is part of our life. Each person builds up for themselves a reality construct, a map of reality. The map is centred on the individual, and everything we know is situated on the map either close to ourselves at the centre or further away. Adults' maps are likely to be large with the adult being more interested in organising the material within the map and puzzling over anomalies than in new areas outside the map.

For many adults, mathematics is remote on their reality map, pushed away from the centre by bad experiences at school, perceived irrelevance, and negative image of the subject. Memories of school mathematics are memories of failure. Many adults are motivated to learn those subjects that not only help them cope with life but also enable them to make sense of their lives and experiences. Mathematics is often perceived as concerned with, at best, meaningless problems about real and material objects but, more often, unreal and meaningless objects.

The reflective learner

The characteristics and aspirations of the learner are crucial to the learning process. This process will be enhanced if the learner becomes explicitly conscious of their own active involvement. The process of constructed knowing requires the weaving of all the learner's life into a recognisable whole. This requires extensive self-reflection in all aspects of their living. In terms of learning mathematics, this self-reflection can be developed and encouraged by the use of learning diaries and mathematical autobiographies. Developing the skills of a reflective practitioner encourages the learner to see themselves as professionals in learning. It allows them to speak and listen to others whilst simultaneously speaking and listening to themselves. This process is a way of knowing and learning that can be used by all from basic education to the highest level of mathematics. This is not just a way of knowing mathematics but also a way to 'make a life'.

Adults decide to return to learning mathematics from necessity or choice. They begin to discover, sometimes slowly and painfully, feelings of greater personal legitimacy. Use of reflection can encourage this but can also cause anxiety as shown in processes of change and conscientisation described in the women's movement (Lea and West 1994). There may be pain in revisiting and reinterpreting earlier mathematics experiences. Revising self-narratives is a way of managing the learning transition and giving it meaning. Telling stories may be an essential of reconstituting the self. Students who are better able to notice and feel their successes will learn more mathematics (Beveridge 1995).

The learner may wish to learn academic mathematics, may understand the framework within which the learning takes place but still find it difficult to learn. Learning in real situations is a very complex business. The tutor cannot presume that the learning that they hope to elicit will actually take place. This will depend on the learner and what they bring with them to the experience. This is determined by the learner. The process can be encouraged by unconditional positive regard and teaching approaches which encourage reflexivity (disciplined reflection) in the learner.

The effect of the pedagogy on the learner

The approach taken by the tutor is critical to how students perceive mathematics and how they learn. The tutor provides the learning scaffold, facilitating the learning and managing the learner's knowledge formation. The tutor, through their pedagogy, can encourage disciplined reflection in the learner and legitimise socially-distributed knowledge allowing more players into the knowledge game. They can show by their approach that they consider there to be no privileged stakeholders and that they believe in open intellectual systems.

In some adult classrooms, the tutor may have little control over the curriculum. Nevertheless they can empower adult learners by respecting their adulthood and their goals, valuing and utilising their background and experience. They can ensure that problems with learning mathematics are not all seen as located in the learner; and can make explicit society's role in their construction.

Democracy must be learned by practice and the tutor must choose to what extent their classroom is autocratic or democratic. Fear of mathematics is not an inherited tendency. It can be created when tutors place too much emphasis on memorisation and the application of rules. It can result when tutors fail to realise the critical connection between academic performance and learners' feelings about themselves, learning and mathematics (Dodd 1992).

Evidence as to what actually happens in the adult mathematics classroom is limited. A British survey of adults learning mathematics on Access courses (Benn and Burton 1993) concluded that students arrived with considerable anxieties often based on bad school experiences. However, the approach taken by tutors was critical in how students came to perceive mathematics and how they succeeded in what was for some a difficult subject, both emotionally and practically. The learners consistently praised tutors for their teaching styles and approaches.

A more in-depth survey into adult mathematics education in three introductory-level courses taught by different tutors in a Canadian community college explored the teaching processes in mathematics education for adults and how they were shaped by social and institutional forces (Nesbit 1995). It found that within the classroom, the tutor was paramount. Almost all decisions about classroom activities were made by tutors whilst the learners' influence was minimal. Decisions were made with little consideration for the needs and interests of the learner though they were described as 'in the student's best interest'. Tutors subtly reinforced the notion that mathematics is a difficult and intrinsically uninteresting subject best tackled by motivation, hard work and repeated practice within a cycle of presentation, practice and assessment. The teaching styles were uniformly teacher-centred, individualistic, competitive didactic and obscured by 'jargon' or complicated language. Knowledge came from the tutor and the textbook, and was based on rule-following and assessment. There was little discussion, debate or regard for difference. Adult learners in the mathematics classroom were socialised into believing that their own experiences, concerns and purposes were of little value. Use of textbooks appeared to lead to being able to follow rules rather than knowing both what to do and why.

A very real distinction was drawn between classroom mathematics and real life mathematics, with learners encouraged to follow textbook rules and ignore their own problem-solving abilities. On the one hand there was the institutional portrayal of mathematics education as part of a system of lifelong learning to provide opportunities to engage in purposeful and systematic learning. On the other hand the curricula and pedagogies were chosen to reflect predominantly vocational concerns and often-outdated notions of appropriate mathematical skills and knowledge. There was no sign of critical pedagogy.

The discrepancies between the findings of the two surveys seem almost irreconcilable. They may partially be accounted for by the different research methodologies or by the different cultural background. It may be that learner perceptions differ fundamentally from researcher perceptions in that learners judge on their likelihood of 'success' in obtaining the qualification, and the researcher in terms of an emancipatory element. It may also be that tutors bring a different style of teaching to Access because of either a freer curriculum or the different expectations of this group of students. Whatever, it does seem that the authoritarian, tutor-centred classroom does still exist in at least some of today's adult mathematics classrooms.

The effect of the curriculum on the learner

For mathematics to have an impact on individuals' lives, it must have not just general meaning but also a personal meaning. The curriculum designed to convey mathematics for its own sake, for the training of the mind, or as an important part of education *per se,* has failed all but a few. Similarly, the curriculum based on strictly utilitarian purposes for a well-educated workforce has not worked, as has been shown by the avoidance of school mathematics in the workplace and the development by different trades of their own ethnomathematics. A successful curriculum needs to see mathematics as part of the learners' attempts to understand their own individual and collective lives and make their lives meaningful. It should be premised on a belief that through mathematics, learners can come closer to understanding the society in which they live, and their own and others' experiences within that society.

The mathematics curriculum is of central importance to the learner. The tutor's pedagogy may be humanistic but if the curriculum is authoritarian and separated, then this may still result in a classroom that imparts banking or reproductive education. A curriculum that divorces mathematics from its social and political context and presents it as a cold impersonal hard-edged subject, will result in, at best, shallow separated learning and, at worst, total alienation.

Chapter Six

Locating the tutor

The tutor and personal goals

Adult educators are often part-time, underpaid and on short-term contracts. Nevertheless, many give more than what is officially required of them because they gain considerable professional satisfaction from working with adults with their high levels of personal motivation. Their goals are likely to include a commitment to give all learners the experience of success and confidence in mathematics and help all learners achieve their own goals whether these be utilitarian, for personal development or for wider social purposes. Some will also seek to develop critical thinkers and active citizens. The beliefs, value system and personal goals of the tutor will affect their pedagogy and it is therefore important that these are made explicit. However, it may be difficult for the tutor to distinguish between their own goals and those of society as expressed through institutional and political pressures.

The tutor and political imperatives

Adult tutors work in further education colleges, adult education institutions, other bodies such as the WEA, university adult education or the universities themselves. Each sector and each institution within a sector is in a rapidly changing climate and has its own purposes, ethos and beliefs influenced by financial and political imperatives. Since 1992, the growth in external controls in funding accreditation and the growing managerialism has led to an increased influence of the institutional context on pedagogic styles. Political imperatives affect funding methodologies, institutional ethos and curriculum design and hence have a profound influence on the tutor.

The tutor and cultural forces

Tutors who see mathematics as fixed, absolute and God-given and mathematics education as context-free will not acknowledge any effect of cultural forces on the curriculum or pedagogy. Those who see either mathematics as a social construct or mathematics teaching as context and culture-dependent may sympathise with feminist theory, influenced by postmodernism, which asserts the need for the tutor to recognise and value fragmentation, difference and diversity in all groups whilst acknowledging their own 'positioned' being. Such tutors may recognise the need to accept the partial knowledge of their own and others' constructed identities. This involves recognising that the standpoint of themselves and the learners, as shaped by their experience of class, race, gender and other socially-defined identities, has powerful implications for pedagogy in that it emphasises the need to make conscious the position not only of the learner but also of the tutor as well.

In this perspective, the tutor needs to both see themselves and be seen by their students as a 'located' being, sensitive to context and situation. This implies that the tutor should recognise the realities of tensions from different histories through an acute sense of difference. This acknowledges the need to validate both difference and conflict but build pedagogy around common goals rather than the denial of difference (Weiler 1995). Tutors need to be aware of and actively fight marginalisation through racism, ethnocentrism and sexism and recognise that discrimination can be discursively constituted through personal and professional taken-for-granted beliefs and practices.

This recognises that present-day provision privileges the learning of some at the expense of others. Within this statement lies a serious problem for many concerned and caring adult educators. It is hard to consider that one of the fundamental tenets of adult education, namely that all learning should be located in student-centred practices, can lead to a concentration on individual and a neglect of social factors. The use of inclusive pluralistic terms like adult learning, adult education or lifelong education can cover and make invisible social collective disadvantage (Rockhill 1995). The notions of student-centred and self-directed learning may need to be re-examined and reconstituted to ensure a genuine recognition by the tutor of the impact of cultural forces on the learners, the tutor themselves and hence on the pedagogy and its outcome.

The tutor and philosophies of mathematics, education and teaching mathematics

Many adult educators 'just happen' into adult education. These tutors are usually not trained adult educators, often not trained teachers and sometimes not trained mathematicians. Each mathematics tutor has some knowledge of the subject but whilst for many this has a base of higher level training in mathematics, some tutors of adults are not mathematicians in the conventional sense. Much numeracy work is taught as adult basic education by tutors who specialise in literacy or have their formal training in other disciplines such as the social sciences. Their knowledge of teaching mathematics is usually practically rather than theoretically based and it is this empirical knowledge that is the essential bridge between their mathematical knowledge, teaching of mathematics, and the teaching of maths to adults. Whilst arguing for a more systematic approach to adult educator training, it is important to recognise that this diversity often brings with it a rich background of content and concepts in other disciplines which can provide sites for examples.

Dominant educational theories affect the tutor. The dominant theory in adult education is progressive (student-centred in a warm, comfortable, user-friendly environment with the tutor as facilitator) and based on the notion of self-direction in learning, with some influence from radical writers such as Freire (1972; 1984) and Lovett (1988). Tutors' theories-in-use may have been gained, for some, consciously through formal education or, for many, informally. This means that for many this knowledge is subconsciously held and never or rarely explicitly recognised.

Research with tutors suggests that few have formal training in adult education and few see the value of this in terms of career progression (Benn 1994). This may be changing in further education where colleges are increasingly demanding a competency-based teaching qualification on appointment. Still, many tutors construct their theories-in-practice from informal staff development such as Access Validating Agency work or research and practitioner bodies such as the Adults Learning Mathematics Research Forum (ALM). The enthusiasm with which ALM was welcomed shows that many adult educators in mathematics feel isolated, with few sources available to aid construction of concepts and theories, examine empirical results of adults learning mathematics or inform the tutor with regard to their role in the teaching process. Many feel that teaching adults mathematics is a rewarding but isolating experience.

The tutor's understanding of the process of teaching and learning mathematics is crucial in the internal construction of the tutor's pedagogical approach. Whether formally trained or not, each tutor will have developed an implicit or explicit personal model of the process of teaching mathematics. These models will be located at some point on a continuum with a narrow instrumental skills-based view and a mastery of facts and skills approach at one end and, at the other, a broader, creative and exploratory view of mathematics as a unified body of knowledge based on understanding. Each tutor will also have developed a personal model of learning. Again there is a continuum with a view of learning as passive reception of knowledge and compliance on one end and the active construction of knowledge and the development of the learner's autonomy at the other.

There is a further major internal force operating on the tutor which is their personal attitude to, and philosophy and beliefs about, reasoning and mathematics. The tutor's attitude to mathematics may vary from enjoyment and interest with a confidence in their own mathematics ability, a good self-concept and a valuing of mathematics to a lack of interest, lack of confidence in their own mathematical ability and a denigration of mathematics as a subject. The attitude to mathematics will undoubtedly affect their pedagogy. So also will their preferred form of reasoning.

Two different philosophies of mathematics useful to the discussions are the absolutist philosophy which views mathematics as an objective, absolute, certain and incorrigible body of knowledge and the fallibilist which views mathematics as a social construct, therefore open to revision and fallible. This can usefully be linked with the concept of separated values and knowledge

(Belenky *et al* 1986) which exclude the personal and laud 'objective' analysis; and connected values and knowledge located in non-judgmental empathy and trust.

Ernest (1995) has constructed a simplified model of the relationship between the tutor's personal philosophy of mathematics, the tutor's value system and the resultant classroom ethos. He suggests that an absolutist philosophy of mathematics combined with separated values will probably lead to a separated vision of mathematics teaching, resulting in a classroom projecting a separated image of mathematics and hence producing separated learning. On the other hand, a fallibilist view of mathematics together with connected values will tend to result in a humanistic and connected view of mathematics teaching which will most likely result in a humanistic classroom where connected learning will occur. Ernest draws into this model two other forces acting on the tutor. First, whilst many mathematics tutors love mathematics *for and because of* its elegance, purity and absolutist features, they are at the same time strongly influenced in their teaching by their connected value systems and their humanist philosophies of teaching and learning. A tutor with an absolutist view of mathematics but with value system based on humanist ideas of connectedness may position themselves in a humanist vision of mathematics teaching. Second, there is another powerful force acting on the tutor, that of the political and economic imperatives of the government, quangos and funding bodies which will normally be filtered through institutional expectations. Tutors with humanistic person-centred views of the classroom, whether with fallibilist or absolutist philosophies, are increasingly being forced by institutional constraints into a more authoritarian, separated 'back-to-basic' classroom. In this case, the articulation of the tutor's personal pedagogy and that of the curriculum will be dislocated, causing potential distress and insecurity to the tutor.

The effect of the learner on the tutor

But this may not be the only conflict. There may be a mismatch between the tutor's connected pedagogy and the expectations of the learner that the tutor will and should use unconnected techniques. Most adult educators will be familiar with the learner who wishes to 'sit at the feet of a master' and receive pre-digested wisdom. Through school experiences of being told that it did not matter that they did not understand, such learners may have built up an expectation of learning that it is not about understanding but about trusting in the teacher and doing what they are told. These learnt patterns of study behaviour can cause severe conflict in a connected adult classroom. Even such experienced and charismatic adult teachers as Carl Rogers write of the difficult outcomes with some learners rejecting this approach entirely and leaving the class (Rogers 1967). It may be tempting for the tutor to reflect the learners' expectations by giving them well-prepared authoritative interesting sessions which, despite a humanistic classroom atmosphere, are, in fact, reverting to the empty vessel or banking approach to education. Here knowledge is constructed and owned by others and transmitted to the student who gratefully receives and stores it. This approach may work well with certain privileged groups in society who just need 'topping up' but has consistently failed marginalised 'others'.

Whatever the approach, however, the commitment and involvement of the adult education tutor is strengthened by the commitment and motivation of the students who turn out after a day's work in all weathers and travel long and /or difficult journeys, despite often being embroiled in serious personal and domestic problems.

The reflective practitioner

The tutor can be more effective in their professional role if they become explicitly conscious of their own active processes and involvement in the teaching situation. This can be facilitated by the adoption by the adult educator of Schon's reflective practitioner approach outlined in Chapter 2 (Schon 1983). This approach allows the individual tutor to step back from their involvement in the teaching and learning process and view it in the context of a wider framework. It encourages the tutor to position themselves in their teaching, be aware of their own standpoint and that of their students and to see the teaching and learning process in a wider historical, social and political framework. In institution after institution and programme after programme, the value of the learner's experience is acclaimed but then ignored as a basis for the design and management of educational activities. The reflective practitioner approach encourages the awareness and critical reflexivity that helps to ensure that this does not happen by a more informed the pedagogy and interpretation or construction of the curriculum. This approach ensures that the tutor is more flexibly and effectively able to cope with the fragmentation, ambiguity and cultural diversity of the postmodern world.

The effect of the curriculum on the tutor

The increasing demand on appointment to further education colleges for competency-based qualifications which tend to focus on practical teaching skills is reshaping tutors' personal philosophies of education. Concentration on these skills leaves little time, space or commitment for consideration of ethical, practical and philosophical issues arising from the educational process or to develop approaches such as that of the reflective practitioner. Ways need to be found to promote engagement, understanding and creativity as well as competence amongst practitioners.

Emancipatory curricula, where they exist, require tutors to think on their feet, to be expert mathematics educators who can recognise and extract mathematics from situations and turn them into curriculum activities. An example of this form of reflective knowing was used in a mathematics classroom in Denmark in the development of a mathematical problem around a comparison of the energy used in the farming of livestock as opposed to crops. The learners used approximating formulae to work out the energy required by the farmer to produce harvested barley, then to convert animals fed with this barley to meat (Bjoerkqvist 1996). They showed that the energy required to produce meat was of the order of five times that required to produce grain: the learners had investigated the critical mathematics of food production. This form of curriculum develops reflective knowing or knowing located in a wider framework. It requires educators educated to teach mathematics and simultaneously think about the process of teaching and the place and use of mathematics in our society in order that they in turn will be able to educate learners to think about these same processes in the course of their learning.

Chapter Seven

Locating the curriculum

The goals of the curriculum

The goals or aims of a particular curriculum for adult education courses will usually be spelt out in publicity and submissions to validating agencies. But behind these explicit statements is an agenda which reflects the role education plays in our society. Education plays at least a treble role in our society. It governs the production and distribution of knowledge in society; acts as an agent of socialisation of the individual into the totality of relationships in society; and is a force for emancipation and growth, both individual and collective (Field 1992). Each role is inter-related with the others, but also acts to some extent in conflict with them. In their essentials, all three roles are present in or education system so curricula are designed to fulfil these aims. What varies from curriculum to curriculum are the relative proportions.

The curriculum and political forces

In the present day, the first two roles have become predominant. The effect of political forces on the mathematics curriculum is considerable and has resulted in recent years in a utilitarian, assessment-driven curriculum based on objectives and competences. The predominant voice in education is that of the New Right which has strongly influenced the design of the National Curriculum and encouraged the development of National Vocational Qualifications and other competency-based qualifications. The current political view of education is that it is for socialising and training.

The curriculum and cultural forces

Our society is characterised by cultural diversity. In order to reflect and incorporate all the different cultures, the curriculum needs take account of the different cognitive styles of students and refer to variations in symbolic systems throughout the ages and in different cultures today (Nickson 1992). Cultural diversity can be accommodated by utilising adults' background education, their use of mathematics whether through work, leisure or domestic activities and their experiences of mathematics as members of different cultures, for example, gender, class, age, minority ethnic group. This could be done through discussion, and hence understanding, of the seemingly self-evident logical truths which form the basic operations of thought and through an exploration of the forces operating upon students and their environment which lead to unquestioned assumptions and attitudes.

Our society is also characterised by structural inequality. To help counteract this, each adult education curriculum needs to seriously engage in equal opportunities through development of policies and procedures and active combating of sexism, racism and classism in the syllabus, pedagogy, text books, materials or assessment. Changes do not happen quickly but are cumulative. The ethos of the tutor and institution needs to ensure that, wherever they occur, stereotyping incidents should be discussed openly and time given to such discussion seen as important. Criticism of discrimination in materials or, more contentiously, pedagogy can increase awareness of stereotyping and hence help to overcome it (Burton 1989).

Ironically, it is the basically liberal and tolerant nature of British culture that sometimes gets in the way, in that we are reluctant to probe or examine learner's cultural assumptions for fear that this will be construed as criticism. The curriculum needs to recognise cultural differences because only then can it incorporate them by commission. If we distinguish, we can begin to eliminate discrimination. If we first recognise difference, then we can learn not to discriminate by omission.

The curriculum and alternative educational paradigms

The dominant approach of adult educators to adult education and mathematics education for adults over recent years reflects at least part of the third role of education in our society that we gave in the introductory paragraph to this chapter: education as a force of emancipation and growth, individual and collective. It has been located in individual growth, reflected in the student-centred, group-orientated, problem-solving, progressive educator approach which views human beings and their growth and development as central. This perspective is very individualistic and does not locate the individual in a political, social and economic matrix nor does it recognise the effect of this matrix on society and the education that society provides.

Interpretations of knowledge are contingent upon how and where the promoters are positioned in society. The extent to which one definition of knowledge is recognised or how far other definitions are resisted depends on the interplay of dominant power structures. The discourse of academic mathematics represents the authorised truth and practices of formal mathematics. The power base of academic mathematics is defended through a cultural capital of values which over time have become entrenched as the 'common sense' against which all eternal values are judged (Preece 1996). Educational discourses, forms and practices have played a significant and powerful role in the maintenance and legitimisation of objective knowledge and scientific rationality. A consequence has been the suppression of 'other' and the acceptance of a white Western male middle class norm.

There is a need for a reconfiguration of adult learning opportunities through increased difference and space for a diversity of voices. The decline of the meta-narrative of academic mathematics and the greater significance of localised ethnomathematics requires the skill of working with difference with a diversity of learners which enable tutors to be explorers rather than preservers of tradition (Edwards and Usher 1995).

The forces of postmodernism with their redefinitions of knowledge may contribute to the reconceptualisation of mathematical knowledge moving from modernistic absolutist to a postmodern constructivist paradigm. Constructivism is a theoretical stance about knowledge, its creation and relationship to the world which sees knowledge as actively constructed by the learner, not passively received from the environment. 'Coming to know' is an adaptive process that organises one's experiential world, rather than discovers an independent pre-existing world outside the mind of the knower.

Ascher (1991) uses the notion of the circle and line to exemplify this concept. She notes the importance in Western cultures of the straight line, flat surface and right angle, arguing that to adults from these cultures these forms are necessary, sensible and proper. For Native Americans, on the other hand, the circle is the fundamental shape and the square is against nature. Each culture sees their geometric form as 'natural' and right. This example suggests that geometric ideas are an integral part of a culture's world view and are hence social and cultural constructs.

In the same way as cultures construct knowledge, individuals are not given knowledge but construct it themselves. Learning or 'coming to know' is the process of adapting one's view of the world as a result of this construction. If learning is seen as the continuous act of making sense and fitting into experience rather than the absorption of preordained mathematical knowledge, then teaching is the provision of opportunity to make sense and encounter constraints and anomalies rather than to convey knowledge. This has a major impact on the curriculum (Jaworski 1991). At a stroke, it exposes the existing cannonisation, exclusion and rarefication of the formal mathematics curriculum and demands an alternative.

There are other ways of examining the paradigms which operate on the curriculum. Four different perspectives of the mathematics curriculum can be identified which have a fundamental effect on the educational outcome (Willis 1995). The first perspective is remedial where what is to be learnt, how it is to be taught and how it is assessed are all seen as given. Here the educational task is seen as providing adults with missing skills, experiences, knowledge attitudes or motivations. The second perspective is non-discriminatory where what is to be learnt is given but how it is to be taught and assessed is not. The task here is to draw upon and extend the learner's experiences, provide a supportive learning environment and more valid assessment opportunities. The third perspective is inclusive where each curriculum is just one of a selection from a wide variety of curricula and therefore neither given nor unchangeable. Here the aim is to provide the learner with curricula which better acknowledge, accommodate, value and reflect their own and their social group's experiences, interests and needs. And lastly, the curriculum can be seen from a socially critical perspective where it is actively implicated in producing and reproducing social inequality, being one of the means by which dominant cultural values and groups interests are maintained. Here the aim is to help individuals to develop a different view of mathematics, understand how they are positioned by mathematics and how to use it in the interests of social justice.

The effect of the tutor on the curriculum

One of the main purposes of education is for socialisation and reproduction of the *status quo*. The implementation of these purposes will lie primarily with the design of the curriculum but the role of the tutor is crucial. From one perspective, the education of adults can be portrayed as an activity taking place within the context of a particular culture within which the practitioners unconsciously transmit their accumulated 'cultural capital' implicit within the curriculum with little reflection, recourse to philosophy or even an overview of the holistic impact of their activities. Adult education practitioners in this view would thus share a culture – one in which they once were the 'empty vessels' – and now filled to the brim, teach it. Adult educators, in sharing this culture, also share meanings. They are tied into a circle of reproduction which not only reproduces the dominant cultural ideas but also reproduces the shared meanings and understanding of the group. They unconsciously abide by the dominant attitudes and values found within their 'social group' and, through interaction with this group, the *status quo* is not only maintained but reinforced. Hence the educational system is a reflection of the dominant culture, holding the values of this culture as important and transmittable.

However, the adult educator is not a passive 'instrument' and will approach the curriculum from their own 'standpoint' including from their own cultural position, with their own philosophy of mathematics and beliefs about teaching and learning. In some situations, the tutor will have considerable autonomy and control over the curriculum. This may result in the tutor subverting the stated or unstated aims and objectives of the curriculum to their own ends either as a conscious or unconscious process. Curricula can bind the tutor closely through assessment regimes but nevertheless particularly in the adult classroom, very little is known of what actually occurs behind the closed door which gives the tutor considerable power.

The effect of the learner on the curriculum

Whether in fact the learner has any effect at all on the curriculum will depend on the aims and beliefs of both the tutor and the curriculum. If both are governed by an absolutist approach to mathematical knowledge and an authoritarian view of teaching, then the learner will be seen as a passive receiver of knowledge transfer and have little impact *per se*. If, however, mathematical knowledge is seen as absolute and unchangeable but the pedagogy as student-centred and humanist, then the content will not be affected by the student body but the approach to teaching will. When mathematical knowledge is seen as constructivist and teaching as humanist, then the learner must be central to both knowledge formation and the pedagogic approach.

Section 3

Understanding adults learning mathematics

Chapter Eight

Numeracy makes a difference

This chapter first outlines several different definitions of numeracy but establishes that, no matter which is used, the level of numeracy in Britain is problematic. It locates the causes of the low level of numeracy amongst adults in the concept of difference – different school experiences, differences located in social structures, different mathematics of different cultures, different mathematics encountered by individuals in different parts of their lives, different mathematics learnt in formal or informal situations. Different goals, different processes, different achievement, different valuation in society's eyes.

What is numeracy?

Numeracy is to mathematics as literacy is to language. C.P. Snow drew attention to the 'two cultures' and the schism between the world of nature and the world of people but he suggested that they also have commonality (Snow 1959). The language of nature, ie mathematics, and the language of people must both be learnt in the context of realistic use in order to sustain motivation and ensure mastery. Our understanding and use of both of these two languages determines the way we think and hence the way we live. However, the social acceptability of each is quite different. Few speak openly, let alone boast of illiteracy, whilst many seem freely able to say that 'I never was any good at maths'. Both languages are powerful tools for description, communication and representation but whilst natural languages such as English are redundant, ambiguous and concrete, mathematics is concise, precise and abstract. We need both.

As with definitions of literacy, definitions of numeracy vary considerably. The definitions of literacy have become more complex in the last fifty or so years to deal with an increasingly globalised world and have moved from the basic concept of functional literacy (using printed and written information in order to function in society). The definitions are now more sophisticated, taking into account the increased literacy demands in this functional sense but also the demand for cultural literacy, scientific literacy, environmental literacy, etc. The definitions of numeracy have also been expanded to take account of our more complex technological society with the advent of calculators and computers, changes in units of measure, and increased statistical information through government agencies and so on. Nevertheless, the emphasis in many curricula is still functional. This is not surprising as the reduction of numeracy to a checklist of coping skills fits well to the present competency-based approach to education and its assessment. However, this emphasis on function, coping and survival focuses on the problems of today and helping the learner to adjust to prevalent social, economic and political situations. This leaves the individual inadequately prepared for either the future or to question and challenge structural inequalities.

The term 'numeracy' appears to have been first used in the report of the Central Advisory Council for Education known as the Crowther Report (1959). It was defined as the mirror image of literacy and the minimum knowledge of mathematical and scientific subjects which any person should possess in order to be considered educated. This covered a broad range of mathematical and scientific understanding including the ability to think quantitatively and avoid statistical fallacies. However, probably the most quoted definition is that given by the Cockcroft Report (1982) which defines numeracy as:

- an at-homeness with numbers and an ability to make use of mathematical skills which enables an individual to cope with the practical demands of everyday life.
- an ability to have some appreciation and understanding of information which is presented in mathematical terms eg graphs, charts, tables, percentage increase and decrease.

This covers wider aspects of numeracy than just basic computational skills, including confidence and critical appreciation. Another practical definition building on this approach is that a numerate person has:

- a sense of number;
- the ability to quantify and estimate;
- skills in measuring;
- usable knowledge of the basic facts;
- the ability to select the appropriate operation to find a solution;
- the ability to use a calculator to perform operations;
- a money-wise sense.

(Frye 1982)

It is probably true, however, that for many, numeracy has a more limited definition along the lines of the ability to perform basic arithmetical operations. This is reflected by ALBSU (1993) who, when defining the numeracy standards for Numberpower, defined the five skill areas of numeracy as the ability to:

- handle cash or other financial transactions;
- use till, calculator or ready reckoner as necessary;
- keep records in numerical or graphical form;
- make and monitor schedules or budgets in order to plan the use of time or money;
- calculate lengths, areas, weights or volumes accurately using appropriate tools *eg*, rulers, calculators, *etc.*

However, the National Curriculum Council (1990) is closer to Cockcroft in stating that numeracy includes the ability to:

- understand and interpret numerical data presented in a range of forms;
- present numerical data also in a range of forms appropriate to different purposes and audiences;
- select and apply numerical methods to problem solving.

The National Council for Vocational Qualifications (NCVQ) (Oats 1990) has defined the core skill area of numeracy in a similar manner around the themes of gathering and processing data; representing and tackling problems and interpreting and presenting data.

The two definitions of numeracy that most closely reflect the approach taken in this book are first that to be numerate is to function effectively mathematically in one's daily life, at home and at work (Willis 1990). As an extension of this,

> To be numerate is more than being able to manipulate numbers, or even being able to 'succeed' in . . . mathematics. Numeracy is a critical awareness which builds bridges between mathematics and the real world, with all its diversity. Being numerate also carries with it a responsibility, of reflecting that critical awareness in one's social practice. So being numerate is being able to situate, interpret, critique, use and perhaps even create maths in context, taking into account all the mathematical as well as social and human messiness which comes with it . . . Unlike mathematics, numeracy doesn't pretend to be objective and value-free.
>
> (Yasukawa and Johnston 1994)

There is a further concept, not prevalent in today's literature but still in the minds of many people, that mathematical ability is a genetic, inherited skill. This leads to the attitude that 'some people are born able to do mathematics, others weren't so shouldn't bother to try'. The premise of this book is that whilst acknowledging that mathematical ability varies across the population, the fundamental problems with learning basic mathematics are not genetically but socially and politically determined. This then raises the question as to whether the responsibility for innumeracy lies with the individual or with society. Later chapters will discuss in depth the contribution of factors such as gender, ethnicity and social class to the seeming lack of mathematical ability of many individuals and argue that cultural forces in society, education and access to education are major contributory factors to levels of numeracy.

This variety of definitions of numeracy leads to similar confusion over the term 'innumeracy'. As with illiteracy, the term is useful but imprecise. Whatever the definition, innumeracy is a problem. As Paulos asserts, 'innumeracy, an inability to deal comfortably with the fundamental notions of number and chance, plagues far too many otherwise knowledgeable citizens' (1988) and similarly for Holfstadter 'for people whose minds go blank when they hear something ending in 'illion', all big numbers are the same, so that exponential explosions make no difference' (1982).

Levels of numeracy amongst adults

The main evidence about levels of adults' numerical abilities in Britain comes from a small number of surveys. The first major one was the survey of adult numeracy levels submitted to the Cockcroft enquiry into the teaching of mathematics in schools and was carried out in 1981 by Gallup (ACACE 1982). Based on a sample of almost 2,900 adults, it found that questions on simple operations such as addition, subtraction, multiplication and division were answered correctly between 64 percent and 88 percent of the time but the questions on railway timetables and rates of inflation were answered correctly less than 60 percent of the time. The level of performance was related to gender, age, socio-economic class and the terminal education age (Evans 1991). As always, such surveys are open to dispute in terms of sample, process and type and structure of questions asked. Nevertheless the survey and consequent report raised serious questions over numeracy levels of the adult population. All subsequent work echoes this concern.

In the late 1980s, a Yorkshire Television advertisement '2+2=5, *If you experience problems with maths write in*' drew over 8,000 replies with problems including helping children with homework, difficulties at work hindering promotion and difficulties in scoring in games such as Scrabble (Hind 1993a).

In October 1990, the BBC launched a series, *A Way with Numbers* (1990). To coincide with this, BSAI released the results of a survey it had commissioned to establish the extent of adult numerical competence and incompetence in Britain. 1,034 detailed interviews were carried out with adults between the ages of 21 and 60. The results shown in Table 2 indicated that there has not been much improvement since Cockcroft (Hind 1993a).

In 1992, a survey of 1,650 21-year-olds was carried out by ALBSU to assess the extent of literacy and numeracy problems. Although most respondents had little difficulty with most of the literacy tasks, difficulties with numeracy were much more widespread (ALBSU 1992). Again the majority of respondents had problems with several of the deemed 'everyday' activities which ranged from calculating change in a shop to extrapolating information from a statistical table. Another survey commissioned by ALBSU and carried out by Gallup in 1993 attempted to establish the use of basic skills including literacy and numeracy in everyday life. 1,060 people were interviewed and of these 13 percent confessed that they avoided activities which involved making numerical calculations (ALBSU 1994).

Numeracy problems and low levels of mathematical qualifications are found in all groups of adults. Even those with higher levels of education are not immune. In the population at large, Cockcroft found that 31.7 percent of his sample had no mathematics qualification; 42.7 percent had CSE grades 2, 3, 4 and 5 or O Level D and E (*ie*, GCSE D, E, F, G equivalent); 25.6 percent had

O Level grades A, B, C or CSE grade 1 (*ie*, GCSE A, B) (Cockcroft 1982: 150). In 1993, 61.5 percent of students in Further Education Colleges were at Stage 1 or Foundation level or below, defined by City and Guilds Numberpower (ALBSU 1993).

Table 2. Numeracy among adults

	SAMPLE QUESTION	RESULTS
MONEY	*How much would it cost to buy a T shirt for for £7.50, a pair of socks for £2 and a pair of jeans for £15?	*12% could not cope with simple addition.
	*Calculate the cost of a dozen chocolate bars at 30p each.	*13% could not cope with simple multiplication.
	*How much change would you get from a £10 note if you had bought something for £3.60?	*11% could not cope with simple subtraction.
	*Divide a bill of £30.35 five ways.	*40% could not cope with division.
MEASUREMENT	*Can you measure from A to B on this diagram?	*15% asked for help in using a ruler.
	*What is the area of a wall 12ft by 8ft?	*39% gave the wrong answer. *20% did not know what was meant by area.
	*Measure out the medicine for a child.	*12% got the wrong dose.
TIME	*What is the latest train that you can catch to make sure you were back at the station to meet someone at 7pm?	*34% could not cope with a simple timetable. *30% had difficulty with the 24 hour clock. *39% had difficulty adding time.
PERCENTAGES	*If you borrowed £6,000 to buy a car and the flat rate of interest was 12% how much would you pay in the first year?	59% could not calculate this.
	* How much is £80 plus VAT?	*47% could not work this out.
GRAPHICAL INFORMATION		*12% could not interpret information from a bar chart. *46% could not interpret a simple line graph. correctly.

Survey undertaken by RESEARCH RESOURCES LTD for BSAI October 1990.

In 1992, the University of Exeter investigated whether the mathematics GCSE requirement for courses such as teacher education and some social sciences is a barrier to higher education for adult returners on Access courses (Benn and Burton 1993). 1,471 adult returners and 253 potential returners completed the questionnaire. The main conclusion was that a sizeable proportion of both groups consider mathematics to be a high barrier for themselves and this is true whether the question is asked directly or indirectly. Respondents express more concern when asked directly if mathematics is a barrier indicating that the very question caused them to recall not only their own experiences and fears of mathematics but also the negative views of mathematics within our society. The potential returners group expressed higher anxiety levels indicating that a significant proportion of interested people may be deterred from returning to study by their fear of mathematics.

Cornelius' research (1992) with new graduates in employment found that, whilst it is probably true that a majority of graduates can handle basic mathematical ideas with confidence and fluency, there exists a sizeable minority who are frightened of mathematics and encounter considerable difficulty with simple mathematical concepts and processes. He also found that undergraduates need help with the basic mathematics required in employers' entrance tests and 1 in 6 had experienced mathematical difficulties at work. He concluded that a lot of mathematics is needed and being used in many contexts in many levels and that there still exists a massive problem even for those who have successfully completed initial education to degree level.

Taken together, these surveys show a widespread problem with numeracy but, even more worrying, a pervasive fear of things mathematical. In the Cockcroft survey, more than half the people originally identified refused even to take part in the study. A similar reaction has been noted in other research. This contrasts with Coben and Thumpson's experience (1995) of asking adults to talk about their mathematics life history. All were eager to take advantage of the opportunity to discuss their fears. Notably this research did not, however, involve a test of mathematical skills. Cockcroft revealed the extent of the feelings of anxiety, helplessness, fear and guilt in people's emotional response to mathematics, their inability to understand simple percentages such as 10 percent tips or sales tax and the common perception that a fall in rate of inflation should cause a fall in prices. These findings are echoed in more recent surveys.

People do cope, but by organising their lives so that they make virtually no use of mathematics. They remove the need for quantitative skills by relying on others, for example, the local plumber relying on his wife to 'do the books', or by constructing strategies, for example, writing cheques or tipping by rounding to the nearest pound. This reflects coping experiences in literacy. People also develop 'folk maths' (Maier 1991), seemingly ignoring school mathematics and using methods and tricks developed by themselves or passed on by colleagues. These rely on mental computation, estimation and algorithms that lend themselves to mental use. An example of this is a worker who had frequent reason to multiply by 7. He multiplied by 3, added the result to itself, then added the original number. Another example is the calculation of 17.5% VAT. A common everyday method is to take the number say £46, move it one place to the right, £4.60, half this number, £2.30, halve again, £1.15, and add all four figures making £54.05. This use of invented methods rather than school mathematics indicates real insecurity about what was learnt in school. For the majority of adults, experiences of school mathematics were traumatic. Hence, if a real life problem is thought to involve mathematics then it is avoided and, rather than use school mathematics, alternative methods are devised.

This avoidance and dislike of mathematics is prevalent throughout society, though as we shall show in later chapters, some sub-groups in society are more affected than others. Even well-qualified individuals are susceptible. This dislike can move into the realms of phobia. Indeed mathematics educators teaching at all levels of mathematics have identified the phenomena of mathsphobia or mathephobia where people's minds actually seize up and they 'panic' when required to perform mathematical computations (Frankenstein 1989; Buxton 1984; Tobias 1978). This was illustrated in the returners survey mentioned earlier (Benn and Burton 1993). When asked their feelings upon finding out that they needed a mathematics qualification to achieve their goal of entry to higher education, these adult students were uninhibited in their responses.

> Dismay! It may prove to be a major stumbling block.
>
> I was not very optimistic, as I have great difficulty with maths and tend to have an aversion to the subject, frankly.
>
> Horrified – because I just cannot seem to do it.
>
> (Benn and Burton 1993)

The implications of low levels of numeracy

Mathematics surround us everywhere in our society. Mathematical and statistical ideas are embedded deeply and subtly in the world around us. The mathematical requirements of different groups are varied but insufficient levels of mathematics for the individual's needs can have major detrimental effects on life chances and participation in society. Requirements are diverse so just a few examples are now given.

Lancaster University's investigation of the numerical skills required by older adults in everyday life identified skills some of which are required by all adults at any age (Withnall 1995: 55). Their classification of numerical skills included:

- financial – mainly concerned with aspects of managing personal finance, planning personal budgets *etc*;
- consumer – the process of deciding on and actually making purchases;
- domestic – perceived as necessary for domestic activities such as cooking;
- technological – to understand and operate home-based modern appliances;

- leisure – concerned with a whole range of activities whether personal hobbies or more complex pursuits involving others;
- citizenship – understanding, interpreting and, if appropriate, responding to statistical and other numerical data available in the public domain;
- voluntary activities – concerned with the requirements of specific roles in local voluntary organisations;
- 'mental notes' – for remembering and visualising numbers and number sequences for a range of purposes.

These requirements are just the tip of the iceberg. Consequences of innumeracy are high for an individual. Access to further education and training may be restricted by a mathematical entry requirement. Similarly, access to jobs, careers, promotion and updating may be curtailed. Many employers use mathematics for recruiting purposes regardless of the actual numerical requirements of the job. The barrier may be the individual's own avoidance of any career, job or course which appears to have a mathematics component. All these restrictions may affect the material and social standing of the individual.

The consequences to the individual of innumeracy can be as serious in non-work activities. The inability to, say, appreciate the effects of various rates of inflation, compare true costs of loans, or estimate unit prices can have immediate and real costs. It can also leave the individual at the mercy of others with seemingly greater knowledge (Dewdney 1993). Recent examples include reprehensible advice to move from state to private pension schemes, to choose endowment over repayment mortgages and even to enter the National Lottery. The applications are by no means just financial. Health issues such as whether to have an operation or have a child inoculated are often based on an assessment of risk.

But there are also social and political consequences of low levels of numeracy. Civic numeracy, which will be discussed in the next chapter, is the tool which citizens use to understand the mathematically-based concepts that arise in major public policy issues such as armament spending, international aid, environmental issues, crime rates, health statistics and social policy. Effective participation in society is limited if the citizenry is not capable of interpreting, producing or questioning statistics. Inadequate levels of numeracy will deter individuals or groups from seeking out evidence and intelligently criticising authority.

Many jobs at all levels require mathematical skills. Much of the language of business is mathematical and mathematics is the language of science and technology. A lack of understanding in the handling of numbers in work can impede an individual's career and also lead to errors which may have serious consequences to their employers, customers or the general public. Research on the production of official statistics, after outlining a series of errors due to simple arithmetic mistakes, concluded that 'Serious errors would certainly occur less often if staff had the ability to recognise figures as implausible and the initiative then to get them sorted out' (Government Statisticians Collective 1979: 144).

In complete contrast to formal mathematics, there is a huge industry built around mathematics and logical challenges. Competence in aspects of numeracy underpinning such popular pursuits as betting, darts, snooker and games of chance, indicate that for many adults it is not mathematical thinking *per se* that is the problem but formal school mathematics. This point will be considered later.

Mathematical thinking is also part of human culture. The interest in recent ideas such as Chaos Theory and the popularity of Stephen Hawking's *A Brief History of Time* (1988) illustrate the importance of mathematical thinking in culture. A rejection of mathematics is a lessening of cultural awareness.

Some explanations for low levels of numeracy

School experience

For many adults their problems with mathematics originate in the school classroom. Memories of school mathematics are memories of failure. The cycle of problems appears to start when the teacher fails to communicate effectively. This leads to failure on the part of the learner to understand what is being taught. The learner is embarrassed to ask for further clarification and the lack of understanding is never remedied. The hierarchical way the subject is taught leads to further misunderstandings and undermining of confidence. Pressure on the learner to gain a mathematical qualification for the next stage of life, whether further schooling or a job, leads to feelings of panic and anger. The learner copes by switching off with the result being no interest or understanding of mathematics (Hind 1993a). The survey of adult returners mentioned earlier illustrates this clearly. When asked to comment whether they had particular problems with their school mathematics, a worrying 80 percent of the sample responded in the affirmative. Many of these students were of the opinion that they had been let down by the school system and reported problems of both lack of understanding and communication with the teacher.

> A teacher who obviously knew his subject but who was unable to communicate the maths to his pupils – I had four years of this.
>
> I believed I could not do maths. I was frightened of maths. The maths teacher was not interested in students like me.
>
> I never understood maths and because the teachers in my school didn't know how to teach maths you were given a load of sums to do and left to get on with it.
>
> (Benn and Burton 1993)

Mathematics staff may have an extra aura of infallibility (Maxwell 1988) but the subject is not seen by students as intrinsic to study and real life in the way that, for example, English is. An authoritarian approach seems more acceptable in mathematics and this is emphasised by the tick/cross marking system. The emphasis on speed and accuracy leads to the appearance of an answer-led rather than problem-led curriculum. Time constraints are a major issue. Learners are expected to look for the right answer and it is assumed that it will be found in less than half an hour. Coaker (1985) notes that the mean time between a teacher asking a question and then providing the answer is 0.9 seconds. The average time for a pupil to formulate an answer is 3 seconds! Communication in the form of discussion between staff and pupils and the pupils themselves does takes time but if mathematics is a language then it is these skills of communication that are fundamental. School mathematics is also perceived as having little relevance to 'real' life. This is reflected in adults' comments, such as:

> Panic! Many of the problems I first encountered in class I hadn't used in my normal everyday life, like ratios, fractions, etc.

Such common bad experiences with school mathematics must have a root in teachers' own experiences, expectations and anxieties about mathematics. Keitel's (1996) work with both primary and secondary school trainee teachers suggests a circuit of aversion to mathematics between teachers and learners. She found that secondary school teachers disliked uncertainty, openness, ambiguity and loss of control so had chosen mathematics as being certain, closed and unambiguous which allowed them to eliminate mistakes, disorder and emotions as far as was possible. They saw the mathematics class as clear, didactic and free of mistakes and were taking this control-orientated teacher-centred approach into their own classrooms. This might well explain some of the school experiences given above.

Despite possessing the required entrance qualification, student primary teachers exhibited a fear that they lacked understanding and ability in mathematics and indeed had confused mental maps of the subject. They appeared not to see any meaning, sense, relationship or connection in their mathematics to themselves or society in general. They heard the words that mathematics is important and even beautiful but did not see this themselves.

Most of the trainee teachers, whether secondary or primary, perceived mathematics as a set of rules. A few saw it as a system of thought but none saw it as formatting social organisation and action in our society. If teachers, when students themselves, saw mathematics as a subject that can be passed without understanding, through blind acceptance and rule-following and as intrinsically meaningless, then it is little wonder that many adults have been incalculated by their school experiences to see mathematics as alien and irrelevant and only 'for other people's problems'.

The relevance of formal mathematics

There is a lack of confidence in those areas of mathematics that do not have every day uses. This is supported by Cockcroft's finding ' . . . that algebra is a source of considerable confusion and negative attitudes . . . in some cases this was because the work has been found difficult to understand . . . and had little point.' (Cockcroft, 1982: 20). It would seem that the features of the curriculum experienced by many adults that contribute to mathematics avoidance include:

- rote calculations which do not increase understanding and are obsolete due to calculators;
- memory dependency, as there is no point in learning the proof of Pythagoras if one does not understand it – and if one does understand it there is no need to learn it;
- unmotivated exercises since spurious applications will not motivate learners;
- authoritarianism in mathematics education where learners are told what they are learning will be useful in the future, and anyway they will understand later;
- tests which assume mathematics can be divided into tiny, separate and distinct compartments.

(Hilton 1980)

These are all very real barriers that inhibit the learning of mathematics in school and sometimes in adult education. But a further barrier exists which is particularly relevant to adults. Many adults are motivated to learn those subjects that not only help them cope with life but also enable them to make sense of their lives and experiences. But mathematics is often perceived as a cold impersonal hard-edged subject concerned with at best meaningless problems about real but material objects but more often unreal and meaningless objects (Noss 1991). It is also taught in a way that divorces it from its social and political context. For example, questions in mathematics relating to 'men digging ditches' have little relevance and are rightly now seldom used. Those concerning the military association of mathematics are prevalent and relevant but raise other issues. Firstly, they are normally presented as abstract, neutral examples whereas in reality the intrinsic issues are controversial and political. Secondly, they are gendered, arguably alienating many women or girls. Lastly, they are dehumanising, stripping social issues of human and moral dimensions. For example, the real life outcomes in human terms of a ballistic question are usually ignored. To remove value judgements from mathematics problems is to dehumanise them. No wonder many individuals and particularly adults do not value a subject which so systematically ignores their intrinsic humanity.

Mathematics of difference

A further reason that so many adults are alienated from mathematics may lie in the construct of mathematics that is presented in the formal curriculum. The content and pedagogic approach has been designed to suit certain sections of society and is hence alienating and remote from others. The split in mathematics in Western culture was laid down in the times of the Egyptians and Greeks and separated scholarly mathematics for the elite from practical mathematics for manual workers. This distinction between mathematics for the different social classes was continued by the Romans with the *trivium* and the *quadrivium* which included mathematics for citizens and a separate practical training for the labourer. With some variation, this differentiated mathematics remained the norm using mathematics as a tool of social stratification and control until the 1944 Education Act.

The advent of the bipartite system of Grammar School and Secondary Modern School continued the division of academic mathematics for the top 20 or so percent and elementary school arithmetic for the rest. This divisiveness has survived the move to a comprehensive system. The academic mathematics still taught in schools ensures a class-based differentiated mathematics curriculum which exercises controls through the higher education and employer recruitment practice of utilising mathematical qualifications as a gatekeeper for further education and employment and hence social standing and advancement (D'Ambrosio 1991).

These divisions are not linked to ability but are structural and a deliberate intention to maintain the *status quo*. Halsey (1992), researching into wider access to higher education but locating his research in access to education at all levels, asserts that educational access is the dependent variable of major social, economic and political forces. He found that though expansion has typically been justified as meritocratic, there is much evidence that ascriptive forces continue to determine the patterns of access among social classes, ethnic and other groups. Public credentials or qualifications are increasingly the mark of modern societies, legitimising inequalities of pay and controlling entry into the labour force. As a result parents seek to convert their own class advantages into enhanced opportunities for their own children. Hence educational expansion leads to a credentialist inflation and access becomes an arena of status struggle between social groups. Access expands but is socially shaped to yield absolute but not relative gains in educational opportunities for traditionally disadvantaged groups. As qualifications in mathematics are determinants of access so mathematics and mathematics education contribute to social inequality.

Ethnomathematics

The role of mathematics as a gatekeeper to the world of work is particularly ironic in the light of modern research which suggests that the mathematics learnt at school is not systematically used in work. Maier's (1991) 'folk' mathematics and D'Ambrosio's (1991) ethnomathematics, focus on common interests, motivations and certain codes and jargons which do not belong to the realm of academic mathematics. The awareness of the existence of the folk or ethnomathematics is crucial for formal adult mathematics education. Adults who have a confused background of school mathematics but have developed strategies within their own folk mathematics may be attending further formal mathematics education because their folk mathematics will no longer allow them to cope with changing circumstances. Mismatches between formal and folk mathematics need to be taken on board and dealt with in an appropriate manner. It is important that learners feel able to bring their own culture into the learning experience and these cultures are acknowledged and respected.

Mathematics is part of the cultural knowledge which all cultures generate but which need not look the same from one cultural group to another. This leads to a fundamental re-examination of many traditional beliefs and theories about mathematics and mathematics education. These will be examined in more depth in Chapter 11. Again suffice to say here that for groups with different cultural backgrounds and expectations, a new approach is needed. Structural discrimination and low expectation of certain groups can be illustrated by the accusation that it is usually black female and working class students who are encouraged to take vocationally-based courses with relatively little academic content by teachers (Giroux 1985). It is unsurprising that alienation from mathematics is structurally as well as individually constructed.

Formal/informal mathematics

Every individual will leave school with at least eleven years of formal mathematics education and many will receive more through further, higher or continuing education. A fundamental question is whether this formal learning can be used in everyday living and whether skills learnt in one context can be transferred to another or whether school mathematics is a separate learning domain with its own rules, purposes and goals which cannot easily or simply be transferred into another activity. If this is the case, then the mathematics used in shopping, work or citizenship duties must be developed by the individual themselves or transmitted by fellow members of the activity group. This mathematics will in turn have its own rules, purposes and goals.

What is at question now is the whole notion of transfer: the ability to apply concepts and skills learned in one context to another context. Our education system is founded on the basic tenet of transferability. This is true across the range of provision from university education to NCVQ core skills. Much adult education also assumes the transferability of the skill of critical enquiry. However, students may learn these skills within the domain (and safety) of the adult education class but may either not think or choose to transfer this ability to real world situations. Oats points out that 'core skills offer the potential for enhancing the transfer of learning to new contexts, not least by making people aware of the skills they possess and require in different contexts. But there has been difficulty in translating this widely-held belief into effective practice' (1990: 11). This is illustrated by the experience of the same numeracy question being asked in different contexts generating markedly different results. There are a growing number of practitioners who have identified this discontinuity between school mathematics and 'folk' mathematics (Harris and Evans 1991; Dowling 1991; Barr 1993; Lave 1988).

However, we need to take care in using the terms folk, out-of-school, local, informal or ethno mathematics. Whereas formal mathematics is a bounded domain of knowledge, these others are an infinite mixture and to talk of them as bodies of knowledge is contentious. They are also not distinct from formal mathematics. The practices, representations and procedures used in formal mathematics will have some bearing on what happens in informal mathematics and *vice versa*. The paradox of this formal/informal dichotomy can be summed up as follows: 'On the one hand, the expression A+B takes its meaning from the situation to which it refers. On the other hand, it derives its mathematical power from divorcing itself from those situations' (Resnick 1986: 30). The power of abstraction lies in helping the individual solve problems easily and efficiently, the problem of abstraction is that it is of no use if the individual does not know how or when to use it. The context-free 'single algorithm for all problems' approach is valued by society for its universality but may be rejected by individuals because of its apparent irrelevance and abstraction.

Being numerate makes a difference

Numeracy consists of being able to make an appropriate response to a wide range of personal, institutional or societal needs. To participate fully in everyday living, adults need the ability to understand broader contexts in which numerical demands are located, to make use of appropriate communication skills, to be able to collect, present and interpret information presented in a variety of mathematical ways and to judge according to the nature of the activity and the desired outcome. People learn best when they are personally involved in the learning experience since learning has to be discovered if it is to have personal significance or make a difference. Considerable research has shown that people do use mathematical concepts in their jobs; some has shown that the mathematics learned formally has not been used. Learning about mathematics is only meaningful if it is accompanied by a pedagogy that raises questions about how it is that students produce meaning and how they become engaged in particular learning situations. This can only be done through a critical pedagogy that illuminates the knowledge, needs and concerns of both individual and group learners. Here the knowledge of numeracy is seen as important, not just for utilitarian or abstract purposes, but as part of students' attempts to understand their own individual and collective lives and to make their lives meaningful.

Through mathematics, students can come closer to understanding the society in which they live, and their own and others' experiences within that society. Greater understanding of the structures and processes of society, and increased mathematical awareness, enable adults to examine critically, challenge and participate. The perspective of mathematics as a historical, social and political construct rather than a neutral and objective one, links the subject to wider considerations of citizenship and social responsibility and empowers citizens to understand and challenge the system in which they find themselves. If the emphasis of mathematics education is extended from individual development and utilitarianism to the fundamental requirement of education for an informed and active citizenship, then not only will this contribute to democracy, but it will also overcome at least the barriers of irrelevancy in mathematics education by focusing on issues of concern to all.

To come full circle in this chapter, mathematics is a form of language that cannot be separated from reading and writing; nor is it any more separate from our effective functioning in society than literacy. Just as our spoken language is more than just the ability to read, so mathematics is more than just the ability to do sums. It is about understanding the significance of number within the social, political and economic framework that is our society. It is not about getting the right answer in a sum but about understanding how operations on data can clarify or obscure reality. It is not about meaningless processes applied to made-up problems but about how awareness of statistical techniques allows the learner to acquire a greater understanding of social issues and critically question public policies.

Through examples based on issues of current social concern, the learner can become actively engaged in the process of critical citizenship. By learning how to investigate and present data round such issues, the learner acquires the power to change others' perceptions as well as their own. The assertion that a critical knowledge of mathematics is fundamental to an active and participatory democracy is explored further in the next chapter.

Chapter Nine

Mathematics for democracy and active citizenship

> The essence of democracy being not passive but active participation by all its citizenship, education in a democratic country must aim at fitting each individual progressively not only for his (*sic*) personal, domestic and vocational duties but, above all for . . . duties of citizenship.
>
> (Ministry of Reconstruction 1919)

Democracy

The term 'democracy' has always been a very contested concept and we shall only touch on this complex area sufficiently to give meaning to our subsequent discussion. Raymond Williams (1976: 82–86) highlights two main traditions of meaning of democracy. One, drawing inspiration from Sophocles, identifies it as the exercise of power by the mass of the people. The other locates it in the selection of representatives of the people through open elections, in freedom of association, expression and personal human rights. Whether direct or through representatives, it is fundamentally government 'by the people'. In Britain 'representative democracy' has come to be regarded as synonymous with 'democracy' but if it is still ultimately 'government by the people' then this exposes a need for an informed public who will intelligently participate in the control of affairs (Fieldhouse and Taylor 1988). Hence democracy involves widespread social and political knowledge and active participation in decision-making by the citizenry. This social and political learning can be, but is not necessarily, emancipatory. Harris (1991) comments:

> participating in decision-making can lead to a reorientation of motivation as individuals are encouraged to accept responsibility for the decisions of the collectivity. Self-regarding may be replaced by other-regarding behaviour, and democracy may act as a form of political education capable of schooling people, through discussion and persuasion, into adopting socially responsible attitudes.

Citizenship

The concept of citizenship underpins that of democracy but, in British society, 'citizenship' is an unfamiliar notion. The Commission on Citizenship (1990) found that the word was not in common use and when used it had a diversity of meanings. The nearest formal definition of a consensus view to emerge was that citizenship involves three elements:

- the civil – the rights necessary for individual freedom;
- the political – the right to participate in the exercise of political power;
- and the social – the right to live the life of a civilised being according to the standards prevailing in the society.

However, most responses to the Commission from participating groups and individuals related to negative freedoms such as the right not to be imprisoned without trial and to responsibilities and duties such as obedience to the law. The Commission reached several conclusions of relevance to this chapter. They argue that citizenship needs to be learnt, that it is not only about rights but also about the everyday participation in our society and that this participation is both a measure and a source of society's success. The challenge to our society is to create ways in which citizens can participate fully and effectively in conditions where all who wish can become actively involved, can understand and participate, can influence, persuade, campaign and 'whistleblow' and be involved in decision-making. The challenge for all educators is to contribute to this vision.

However, in our society there are major impediments to this version of citizenship and/or active participation. The foremost issue is whether citizenship is a matter of rights or duties. In particular, the previously unquestioned British welfare state commitment to social rights has now been at least partially displaced by an emphasis on social duties (Roche 1992). A government which views citizenship as primarily about duties may not necessarily wish to develop or support an education system which concentrates on rights and active participation. Lack of knowledge is a serious impediment to full citizenship. For example, the questions of right *to* citizenship and rights *of* citizenship are not clear. Structural inequalities and social disadvantage also restrict active citizenship.

Many social factors such as poverty, ill-health, gender, race or age may disadvantage parts of the population and prevent their participation. The Commission noted these barriers but focused mainly on encouraging citizens to work in a voluntary capacity despite noting their concern that 'voluntary effort is being used to compensate for the deficiencies of the public service' (p.33). They found that many people do contribute voluntarily 'to the common good through the participation in, and the exercise of, civic duty' (p.8). For example, nearly half the public had been involved in fund-raising and a third had helped to organise activities.

Though there has been a steady decline in membership of political parties, there has been a substantial increase in the numbers joining voluntary organisations including environmental groups like Friends of the Earth and Greenpeace; conservation bodies such as the National Trust; pressure groups; Neighbourhood Watch Schemes; women's groups; support for the statutory sector *eg*, the social services by Meals on Wheels, the education service through playgroup associations and school governorships; churches; the trade unions – the list could go on and on. These volunteers must have the skills required to perform their duties. Adult education is of course only a very marginal activity but we will now examine whether it can aid this active participation and help to counteract the forces that hinder participation.

Adult education

The traditional link between adult education and participatory democracy has been 'education for citizenship' and the equipping of individuals and/or groups to take a more active and effective part in running or changing society. This can be achieved through increased participation at different levels of society from local to national and in the kind of organisations, associations and groups outlined above.

Writers such as Fieldhouse (1985, 1988) have written extensively about the means by which adult education and particularly that provided by the universities can contribute to democracy. In this liberal and liberating tradition, it is the approach to the teaching and learning process that is crucial. Students can acquire a critical approach to authority and a capacity to distinguish matters of fact from those of opinion. Provided that this does not result in the student being so constrained by alternative views that they are reduced to sterile neutrality, then the skills learnt can be transformed through praxis – action based on theory – to more active and effective participation in society.

However, this pinpoints an unfortunate schism in adult education. Many writers and theorists assert that this collective form of adult education for emancipation and citizenship 'is regarded as less acceptable or valuable than either a blander education for individual personal self-fulfilment, or the new modes of continuing education intended to retrain and refit people for their economic function' (Fieldhouse 1985: 44), whilst themselves showing the inverse preference. But it need not be an either/or; it could be a both/and. It may be that fundamental changes in society are unlikely to be brought about by collective action alone and individual self-development can be seen as a means of reducing the inequalities that limit opportunities for participation. Individuals' perceptions of themselves and their false consciousness can be changed through individual development and perspective transformation.

It is similarly a shame to denigrate work-related learning as not contributing to participation if for no other reason than that this is probably the most well-attended branch of adult education. The critical approach deemed so central to the liberal adult tradition described in Chapter 1 can as valuably be applied to all forms of adult education no matter what the subject studied. We will return to this point later in this chapter.

There is, however, a strong caveat to any discussion of adult education's contribution to a more active citizenry. The Commission on Citizenship found that structural inequality impedes participation, and citizenship has to be learned. For the young this learning takes place in the school. But here it is likely that the many children who are alienated from school by its reflection of middle-class, ethnocentric, gendered curriculum see citizenship like adult education as 'for other people'. Duke (1992) expresses links between education, learning and citizenship in the following rather depressing diagrams. The innocent and hopeful or perhaps gullible view of the link between education, learning and active citizenship is expresses as follows:

education → learning → active citizenship → impact on the society and state

The less innocent view illustrates the tenuous link between education and learning for many in our society and the institutionalisation of disadvantage. Crudely the top diagram represents the middle classes and the bottom the educational underachievers from other socio-economic and ethnic groups.

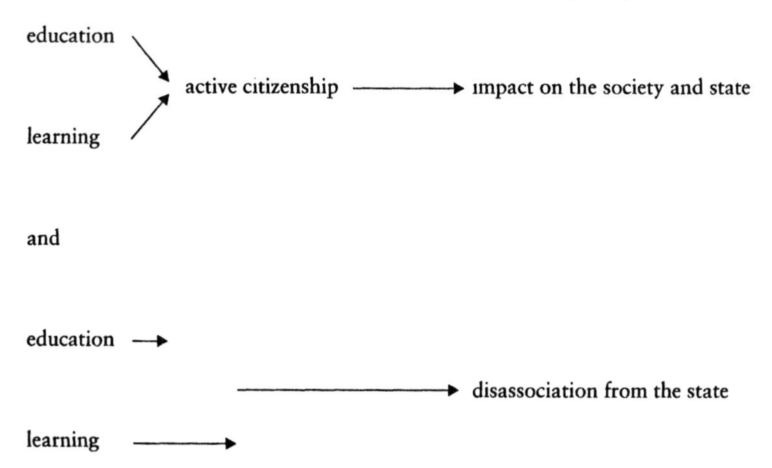

The commitment of many adult educators to increase the number of people who actively participate in a democratic society is rather stymied when it becomes clear that, as was shown in Chapter 2, those who participate in formal adult education are *already* from an active social minority whereas those do not participate in society are typically non-adult education joiners. There are no easy answers to this fundamental dilemma. Whilst adult educators can encourage the enhancement of current students' contribution to society, they will not increase the numbers of active participants in society in any substantive way if participation in provision is not extended. This may be possible through outreach and special targeted programmes. But if these groups are attracted into adult education, it is crucial not to repeat the pattern of alienation. Citizenship has to be learned like any other skill but it will be learned not through the formal curriculum but through positive experiences of participation. Participatory democracy is learned through practice and therefore the adult education experience should itself be an experience of participatory democracy.

In this way it can be an affective as well as cognitive learning experience that both citizenship and adult education are 'for us' and not just 'for other people'.

As we said earlier, one of the justifications for the existence of adult education is education for citizenship, particularly now that all adults have to adapt to life in an increasingly information-laden society. Paradoxically, adult education programmes seldom actually address students' statistical literacy. This is a critical but neglected area of provision.

Mathematics

Active participation in society was defined earlier as allowing those who wish to become actively involved, to understand and participate, to influence, persuade, campaign and 'whistleblow' and be involved in decision-making. If the teaching and learning of mathematics contributes to this form of participation then it has a political focus. This leads to a discussion of the political aims, whether conscious or unconscious, of those who design the mathematics curriculum, its role as either conformist for social reproduction or catalyst for social change and the issues this raises for the teaching of adults.

Active citizenship involves numeracy

Voluntary bodies, pressure groups and women's organisations may require citizens to produce or seek out data then analyse it and understand the context where it was produced. Examples are:

- lead pollution levels in the area around local schools for a campaign against motorway development;
- information gathering by a campaign for secure accommodation for single people;
- investigation of changes in women's employment patterns to inform local authority adult education provision.

Hence certain mathematical skills are needed for critical citizenship and include

- knowing about official information sources and how to access them;
- how to obtain information produced but not published;
- methods for the production of information at a small scale level in the community;
- and the interpretation of information from other sources or one's own research.

(Evans 1990)

An important skill required in the struggle for critical citizenship is access to and a grasp of official statistics from which we obtain most of our information about government spending, unemployment, poverty and so on. This access takes place mainly through the media (one has only to think of television news) but in the context of the current political, social and economic climate. These factors have an immense implication for the accuracy of information disseminated.

Sadly one of the crucial factors in the debasement of political life is the way statistics are now manipulated by politicians, lobbyists and special interest groups to provide supposedly objective evidence for an argument. A stinging indictment of the politicisation of official statistics was made by Murial Nisseel the founder editor *of Social Trends* in the official commemoration of the publication's silver jubilee in 1995. Manipulations include the mystification of trends through the publishing of claims without reference to the full details of assumptions which are necessary to interpret them; frequent changing of crucial definitions (for example, unemployment figures); and ceasing to publish certain statistics (for example, the number living in poverty).

To combat these machinations and help empower citizen groups, Evans (1992), using the analogy of 'barefoot' doctors and lawyers or, more familiarly in Western societies, paramedics, identifies the need for not just a numerate citizenry but also a cadre of 'barefoot' statisticians or para-statisticians. These would not be trained mathematicians or statisticians but individuals who could contribute to their own community groups their ability to handle the data requirements of the group and report the results of any searches or investigations in terms that are clear and comprehensible. Again the crucial role would be the production, accessing or presentation of data. An imaginative idea.

Thorstad (1992) identified school governors as a prime example of citizens who work responsibly and without pay on behalf of the community. Since the introduction of Local Management in Schools (LMS), governors in the state system have assumed financial responsibility for the running of their schools. She concludes that one of the stated aims of LMS, to increase local participation in the running of schools on the grounds of democracy and efficiency, may not be achieved for reasons linked to mathematical ability and confidence. Some people are deterred from standing for governorship in the first place due to lack of knowledge or confidence in financial matters, and even those who are elected may be making decisions on shaky ground due to similar inadequacies.

The numerical skills identified as being of most use to a governor were the ability to: follow an argument that includes (especially large) numbers; do a quick estimation; check other people's calculation; and calculate accurately with speed and agility but using a calculator. This is a mismatch with numeracy practice encouraged at school. Adults were insecure with mathematical skills half-remembered from school or informally learnt as an adult or a confusion of the two. The result was that some non-specialist governors, including parents, did not take an active part in crucial debates or were being asked to rubber-stamp financial decisions made by the financial subcommittee.

Another study into active citizenship investigated the numeracy issues raised by the introduction of the Council Tax (Hind 1993b). The results supported the research quoted in Chapter 8 that between 10 and 20 percent of the adult population have problems with basic mathematics. She found that the resultant inability to interpret numerical information led to a lack of knowledge of new developments such as the Council Tax and a failure to understand its implementation. This meant that the citizen did not have the requisite information to make decisions about, for example, tax payments, the fairness or otherwise of the tax or how to claim for benefits or discounts. This affects the ability of the citizen to operate effectively in a democratic society. The other side of this coin is the responsibility of government to take mathematical limitations into account and provide information in an accurate but easily assimilated and accessible form.

Fostering critical awareness and democratic citizenship for adults through mathematics requires questioning and decision-making, discussion, permitted conflict of opinion and views, challenging of authority, and negotiation. Hence the curriculum must include these components, and materials should include socially-relevant projects, authentic social statistics, accommodate social and cultural diversity and use local cultural resources (Ernest 1990).

However, the numeracy curriculum is currently constructed around the immediate personal or work related curriculum of the individual learner or based on the school mathematics curriculum. It could be extended, as has been done with literacy, to integrate numeracy skills with issues of public concern such as school budgets or new tax proposals. However, adults' expressed needs should not be ignored and any widening of the curriculum should not replace the instrumental goals and self-development requirements of the learner but enhance these. Adults can be encouraged to recognise and value the mathematics learning that takes place in all facets of their everyday life. The role of the adult as citizen, in addition to worker, can provide a wealth of suitable material for accessible everyday 'really useful knowledge'.

A critical approach to mathematics teaching through a statistics literacy course could encourage learners to pose their own questions such as:

- what mathematical questions arise out of this situation?
- what mathematics is being used, or could be used, in this context?
- which groups in the community are affected by the circumstances described?
- which groups are likely to benefit from the use of mathematics in this context?
- could you look at the questions in a different way? Would this produce different answers?
- are there important factors that have been ignored?
- is there any information not given here which might help you answer your question?
- are the mathematical measures used credible?
- what is the quality of the design?
- what are the allowed inferences?
- are the conclusions warranted by the data?

(Barnes, Johnston and Yasukawa 1995; Gal 1996)

The data for critical explorations of important social issues could come from newspapers, official Government statistics or the newsletter of the Radical Statistics Group (RSG). The Winter 1995 edition of the RSG's Newsletter contains useful starting points for investigations such as 'The unofficial guide to official health statistics', 'The Department of the Environment's index of local conditions: don't touch it' and 'Retiring into poverty'. A statistics literacy course could aim to convey basic knowledge such as an understanding of terms, like average, percentages, graphs, and the logic behind certain concepts such as why averages are used. It could result in the learners constructing a critical list of questions along the lines of the ones listed above. Most importantly, it could aim to encourage learners to think statistically and appreciate that the application of statistics is valuable (Gal 1996).

Can mathematical awareness contribute to political awareness?

There is no consensus between mathematics educators on this matter. Some argue that mathematics is antithetical to political awareness because it necessarily operates on the suppression of precisely those elements and functionings that need to be in sharp focus when thinking about the political (Pimm 1990). For example, in any ballistics question in mathematics, political thought would emphasise ethics, values and morality or the results of success or failure of the strike. On the other hand, mathematical thought would of necessity emphasise the symbol manipulation.

Examples and questions on percentages sometimes relate to unequal distribution of wage rises in small firms. Politically this raises issues of ideology. Mathematically, teachers and students alike suppress the meaning and concentrate on the symbol manipulation. Fluency is encouraged at the expense of social and political meaning because what matter are the skills and the resultant qualifications. What is ignored is the consequent depoliticisation of the student and teacher. Or, more cynically, may be not ignored but encouraged to form a more obedient and compliant citizenry.

A counter-argument may be made that mathematical training may, by this very factor of 'de-meaning' situations, develop in the individual a free thinking, liberated from the dominant world views and hence free to challenge social injustice. Bertrand Russell is an example of such a politically-aware mathematician. Writers from Plato onwards have seen a mathematical education as contributing actively to a political training by providing a rigorous background in rational thought, logical thinking, and problem-solving abilities. They see mathematics in itself and for its own sake engendering the educated person (Fauvel 1990). This in its turn is countered by arguments that the requirements for assessing a problem critically are epistemological, not logical, in character and hence academic mathematics gives no guarantee or cognitive basis for critical thinking (McPeck 1981). There may not be a real contradiction between these differing views when it is recognised that the one may be particularly apposite for mathematical learning for the masses with the other only applying to the elite in our society.

Politics and the mathematics curriculum

Mathematical problems centred around citizen issues such as school budgets, the Council Tax or unemployment figures can lead to accusations of a non-neutral curriculum that dispenses propaganda. This is a serious matter and the arguments as to whether or not there is a political dimension to mathematics education need to be examined carefully.

Mathematics has the reputation of being the most value-free discipline. Some mathematics educators would argue that their job is simply presenting the facts, teaching the skills and hence avoiding opinions and values. However, if there is a political dimension to mathematics education, then by not identifying the opinions and values hidden in the data, these are then closed to critical enquiry in the approach advocated earlier in the liberal tradition. Independent judgement is not developed and hence active and effective participation limited.

Radical educators such as Freire (1972) strongly argue that education is never neutral and that non-controversial education is always political in as much as it implicitly supports the *status quo*. This is supported by Jenny Maxwell's research which exposed the 'hidden messages' in mathematical examples used in British schools (1991). She first lists some blatantly political examples of indoctrination from foreign cultures; guerrillas helping peasants in the fields from Mozambique; freedom fighters in Tanzania; agriculture and military applications from China. As outsiders, it is very easy for us to see the hegemonic forces at work in these examples but we feel sure it does not happen here . . . She then made up a small collection of questions on percentages based on textbook questions. Questions included class-bias, gender-bias, and racism but in standard mathematical format. Of her sample of 25 teachers, just under a half commented on these biases and thought the questions 'very establishment'. The rest did not find them so or did not consider this relevant to their teaching of mathematics. After further discussions, about a quarter of the teachers remained clearly of the opinion that mathematics education is and should be politically neutral. This was despite the pervasive and unanimous attitude of guilt and apology whenever a teacher felt he or she was questioning the norms of society. Four expressed fear of being thought leftist but none were anxious about upholding the values of the right. A small but illuminating piece of research. Some of her cohort moved through a perspective transformation and came to realise that 'you may end up putting forward social views that you disagree with'.

The imbalance of questions in mathematics education generally is illustrated in the emphasis on mortgages, investment and interest rather than, say, social security payments. It would seem that the poor are being asked to struggle through the problems of the well-off but not *vice versa*.

Some teachers of mathematics educators are trying to raise the awareness of their students to this political dimension and an interesting example of this comes from South Africa. Chris Breen, at the University of Cape Town, asks his students to comment on three views of school mathematics written from different political perspectives (1990). The technicist view addresses the applications of mathematics in a modern society; the esoteric view argues the beauty and elegance of higher mathematics but denigrates school mathematics; and the emancipatory view sees mathematics as a way of understanding and combating repression by the oppressors. This practical example illustrates the strength of applying the ideas of a liberal education outlined earlier to a 'vocational' discipline. Breen's tentative results support the assertion that giving students access to arguments but then leaving them to make up their own minds, enables them to learn to challenge existing prejudices, both their own and others', and so develop an independent judgement.

Critical citizenship and the design of the mathematics curriculum

To understand whether critical citizenship is seen by the state as a desirable outcome of mathematics education, it is helpful to examine the aims of the mathematics curriculum within the context of the range of interest groups at work in educational policy in Britain today. We shall take as our starting point Paul Ernest's analysis of mathematics education in the school curriculum (1992) which builds on the early work of Raymond Williams on education (1961) and then extend this to adult education.

Ernest identifies five interest groups in education. First there are the *industrial trainers* whose aims in education are utilitarian and concerned with training a workforce in basic skills. The *technical pragmatists* are also concerned with the pragmatic skills for a well-trained workforce but extend their concept of education beyond the 'back-to-basics' of the industrial trainers to information technology capabilities together with communication and problem solving skills. The *old humanists* are interested in the transmission of cultural heritage and knowledge for its own sake. In mathematics education, they are interested in pure mathematics for its beauty and elegance not its utilitarian worth. The *progressive educators* grew out of the progressive tradition and believe in education for individual self-development. In mathematics education, this manifests itself in a belief in student-centred learning through problem-solving in a supportive environment. The *public educators* are radicals concerned with democracy and social justice. They see education as the means of extending participation in all aspects of a democratic society. In mathematics education, this group sees the subject as emancipatory, giving students

> the confidence to pose problems, initiate investigations and autonomous projects, to critically examine and question the use of mathematics and statistics in our increasingly mathematised society, combating the mathematical mystification prevalent in the treatment of social and political issues. The outcome should be individuals who are more able to take control of their lives, more able to fully participate in the economic life and democratic

> decision-making in modern society and, ultimately, able to facilitate social change to a more just society.
>
> (Ernest 1992: 36)

In the development of the National Curriculum, Ernest argues that the public educators and the progressive educators have had little or no impact, whilst the industrial trainers have dominated the successful alliance between the other three groups. The implications of this on the school curriculum are immense and have resulted in a utilitarian, assessment-driven curriculum based on objectives and competences. It can be argued that the National Curriculum is about testing and grading *per se* and what is tested is of less than secondary importance (Noss 1992). If this is so then this lack of concern for content indicates the government's view of education ie that it is more a socialising process than a training one. It also means a curriculum dedicated to the preservation of the *status quo*. The return to an uncritical dependency on the teacher for the 'right' answer is intended to lead to the development of an uncritical reliance on authority in society in general and hence a passive citizenry (Lerman 1992). This analysis suggests that critical citizenship is currently not seen by the state as a desirable outcome of mathematics education.

Mathematics in its political and social context

All of this has direct implication to adults learning mathematics. Many adults study mathematics to obtain a qualification directly linked to the National Curriculum. Others are interested in National Vocational Qualifications which come from the same stable. The dominant approach to adult education and adults' mathematics education over recent years has been the student-centred, group-orientated, problem-solving, progressive educator approach which views human beings and their growth and development as central. This perspective is very individualistic and does not locate the individual in a political, social and economic matrix nor does it recognise the effect of this matrix on society and the education that society provides. The process of learning and the development of the individual is seen as equally, if not more, critical than the knowledge itself. Although this approach to teaching and learning is different from that of the industrial trainers and technical pragmatists, all three groups share the common perspective that mathematical knowledge is certain, neutral and value-free. This would seem to imply that adult educators, with this humanist, progressive approach based on the ideas of Dewey and Rogers, may well feel comfortable, or at least not too uncomfortable, continuing to teach in their existing student-centred way but towards the new goal of competency-driven education.

The popular learning technology of Knowles (see for example 1980, 1984) with its language of experiential, student-centred learning through negotiated contracts is easily adapted to a curriculum based on work and everyday mathematical skills for individuals through competency-based assessment. So current adult education practice may have adapted, or be adapting, without too much difficulty to the hegemonic approach of the dominant industrial trainers.

The concentration by the industrial trainers and progressives on mathematics as a set of skills which will improve career or job prospects or increase personal development and feelings of self-worth, may not contribute to the active participation we have been discussing. The reason for this is located in the common belief of the two approaches that mathematics is absolute, neutral and value-free which leads to a curriculum which ignores the relationship between mathematics and the world it inhabits and is blind to the ideological dimension of the knowledge it transmits. This, public educators assert, cannot lead to active and critical citizenship. If this is our goal, we must look elsewhere.

To investigate the relationship between education and active and critical participation in the democratic process, it is interesting to examine the ideas expressed in Freire's *Pedagogy of the Oppressed* (1972). Though written as a reflection on his literacy work with the oppressed people of Brazil, it can also be seen as relevant to numeracy in our own culture. Freire termed the technical educator's approach to education as 'banking', the depositing of information by the knowledgeable 'teachers' or 'authority' into the 'empty vessels' of the passive students. The students' only action is to receive, file and store the deposits. They are cut off from creativity and enquiry and hence knowledge and action.

Freire's alternative to education for oppression is education for liberation through praxis, 'the action and reflection of men (*sic*) upon their world in order to transform it'. This occurs through conscientisation, the process through which people become aware of how their experiences are structured and conditioned, of the forces acting upon them through the social, political and economic culture in which they live, and aware of the needs of different interest groups in society. In mathematics education, conscientisation is the process where students become aware of the relationship between mathematics and society and how this is related to their own condition, situation and development. These ideas, though not usually applied to mathematics education, may contain the basis for mathematics for active citizenship in the British context.

Most views presented to students, as we illustrated earlier with Jenny Maxwell's hidden messages, legitimate the existing social order. This excludes ideas of change, experience of conflict and the creation by the learner of their own set of values. Education for active involvement needs to counteract the dominant paradigms by presenting other competing world views. Through critical analysis and reflection, the student can be encouraged to recognise and evaluate different paradigms and through discussion of controversial issues, develop frameworks, concepts and approaches. This transfers knowledge and hence power from the elite, into the hands of the people. In this process, mathematics is not seen as valuable in itself but as a tool, a library of information and skills, to be called on as needed when and only when the problem requires it. Critical thinking is crucial and can be developed through problem-posing and problem-generation.

There are several stages to this approach. First identification, the locating oneself within a culture. The learner needs to understand the cultural constraints operating within society but also, importantly, on themselves. They can locate themselves in this framework of forces by a growing realisation of the effects of, for example, sexism, racism and class conflict on their individual experiences of learning mathematics and the larger context in which these experiences take place. Through this raised awareness, the learner can come to a better understanding of the causes of mathematical anxiety and their own, perhaps troubled, relationship with mathematics. Next the learner must become genuinely engage in some form of mathematics. This can take place through the collective generation of problems of interest and hence motivation to the learner. This collective approach can take place in the formal learning situation through dialogue and group work. Then comes the objectification of the problem (standing back and critically reflecting upon the purposes and consequences of studying this problem in relation to wider values). For example, the group may choose to study armament spending. To ensure a critical approach, all data sources are named and data from rival sources is provided. Differences are analysed in terms of, for example, differing interest groups. Critical judgements can be made on the reliability of data and conclusions drawn about the use of data in this controversial issue. This approach generates an awareness of the social and political responsibility of mathematics and locates mathematics in its social and political context. Here statistical data is not used as a vehicle for learning mathematical techniques. Rather the data and questioning this data is used to understand and change assumptions about issues.

This is the start of a critical mathematics curriculum where a critical understanding of numerical data prompts us to question unchallenged assumptions about how society is structured and enables us to act from a more informed position on societal structures and process (Frankenstein 1990; Abraham and Bibby 1992). As an emancipatory process it is very powerful in raising awareness that mathematics belongs to everyone; that the learner and teacher are engaged *together* in the learning and doing of mathematics and the world at large is accessible to analysis, criticism and transformation (Lerman 1992). These attributes contribute usefully to the requirements for active participation and critical citizenship.

Issues arising from this approach

The approaches to mathematics education outlined in the last section would arguably lead to a more discriminating critical citizenry with an increased capacity to take an active part in society if they so wish. However, Abraham and Bibby (1992) give an example of a mathematics question set in 1986 which looked at growths and comparisons in military spending and asked for military spending per head of population and comments on the results. They list the reactions from the more conservative British press which complained, amongst other things, of propaganda in the classroom. As a result of these complaints, it was decided that examination boards should in future vet mathematical papers for political content. 'Political' here clearly means 'anti-establishment'.

Any introduction of curriculum that critically debate the relationship between mathematics and society can expect this kind of response from those groups whose interests are being served by the present system. But adult education is a more marginal activity with considerably greater freedom of action, or at least was until the 1992 Education Acts tightened the purse strings, located much provision under the auspices of the FEFC and linked funding primarily to qualifications. In an assessed course, the prescription of the syllabus almost inevitably limits freedom of content. Even if brave mathematics adult educators risk bringing the wrath of the establishment down on their head and do change their approach, as was suggested earlier little work has been done on establishing whether a critical approach is a transferable skill or whether the learners would even wish to transfer it. Increased critical awareness is not always comfortable and may be contained in a compartment of the learner's life. If the learner does use these skills in a wider critique of society, this may lead to frustration and anger when confronted by forces beyond their power to change. This is not to argue against this more emancipatory form of education, far from it, but educators need to be aware of these potential outcomes. A realistic assessment of the outcomes of any change of teaching are that individuals might be more frustrated in their new state of consciousness but they may also be more empowered to join attempts to change to a more just society. The net result in our consensual society will not be revolution but might lead to a slight shift in the political climate towards a more democratic and participatory society.

This chapter has attempted to explore the complex web of factors linking citizenship and adults learning mathematics. There are no easy answers but the questions are worth asking for the sake of a more just society. What is clear is that without a firm foundation of mathematics, adults will be barred from full participation, and hence mathematics contributes to democracy.

Chapter Ten

The discourse of mathematics

In this chapter, the varied discourses of those involved in the process of adults learning mathematics are investigated. The context is set by examining firstly the dominant and less powerful voices in society and secondly the forces acting upon the discourse of adult educators. The different discourses in mathematics itself are briefly explored before looking at the idea that the study of mathematics can be seen as the introduction of the student to the discourse of mathematics. The remainder of the chapter concentrates on mathematics as a language and the importance for the student of being able to read, write and talk fluently in this language.

Silent murmurings

> rediscover the silent murmuring, the inexhaustible speech that emanates from within the voice that one hears . . .
>
> (Foucault 1986: 28)

A discourse consists of a loose-knit collection of concepts, terms, assumptions, explanatory principles, rules of argument and background knowledge which are shared amongst the members of that discourse community (Northedge 1994). 'Discourse' is used in this book to emphasise the social nature of reading and writing practices. It can be defined in the following way:

> . . . the term points to the fact that social institutions produce specific ways of writing or talking about certain areas of social life which are related to the place and nature of that institution. That is to say, in relation to certain areas of social life which are of particular significance to an institution, the institution will produce a set of statements which largely define, describe, delimit and circumscribe what is possible and impossible to say with respect to that area, and therefore how it is to be talked and written about.
>
> (Kress 1985: 139)

A discourse has the power to create reality by naming that reality and giving it meaning. It is not just a matter of words but real power, with the discourse able to deem what is 'real' and 'true' and hence what is included and what is excluded. What is not named may not even be noticed. The result can be that the social and cultural life of some groups in our society go un-named and un-noticed, hence the 'silent murmurings'. We all belong to a multiplicity of discourses, some informal such as gardening or tennis, some more specialised and elaborate such as law or medicine. Some formal discourses require considerable training to enter but bestow corresponding prestige and status. The discourses we are interested in, both in this chapter and throughout the book, are those of the dominant groups in society (white middle-class middle-aged males?), those of groups less well-heard in our society (including women, ethnic minorities, the disabled), the discourse of mathematics itself, that of mathematics education and that of adult education. Our main interest is in the adult who wishes to use or learn mathematics by acquiring, through study, the discourse of academic mathematics and hence a facility with the language of mathematics.

Notions of common sense and rationality are expressed by the dominant culture through culturally-specific discourses. The dominant rationality in society is based on narrowly-defined boundaries measured against particular norms. This ignores alternative discourses, languages and practices. It also ignores power relations and influences of historical circumstances on the different discourses, their practices and constructions of reality (Preece 1995). Hence certain patterns of human development become 'natural' or the manifestation of progress whilst other cultural values and forms of knowledge are not recognised. The narrative of development forces convergence to the 'same' because difference is either marginalised or treated as a threatening 'other' (Usher 1995). There are alternative realities, truths and discourses which are not recognised by dominant cultures causing the marginalisation of certain groups in society.

Within mathematics and mathematics education in society in general, postmodern thought postulates alternative discourses but again some of these are not valued or represented in Western academic discourse. To hear these discourses, we must 'rediscover the silent murmuring'. Postmodernist deconstruction echoes this desire for rediscovery by focusing on that which may be missing from or hidden in the text and by celebrating diversity, a plurality of perspectives and the partiality of all knowing.

But what forces ensure the dominance of the 'grand narrative' of modernity? Although the lives of many people are best expressed as location in a multiplicity of narratives and hence of a multiplicity of identities rather than location in an underpinning grand story, the need to tell one grand story continues to be strong and an attempt to maintain or recreate a single, simple and unambiguous narrative and identity is common. There is a dilemma between the security of a dominant narrative and the range and diversity of possibilities that are opened up by the alternative realities. There are in addition serious critiques of this 'opening up'. Edwards and Usher (1994) suggest that fragmentation may disempower in relation to a unified intervention (eg, by the state). Westwood (1992) writes that far from generating a radical vision, postmodern politics may be deradicalising and domesticating and that it adopts, rather than provides an alternative to, the language of the right.

Nevertheless it is clear that within the construction of discourses lies power and the ownership of knowledge. In this and later chapters we will strive to rediscover the 'silent murmurings' of those whom academic mathematics has so conspicuously failed. To do this, we must first recognise and analyse further the power of language in our own and others' discourses. We shall start by examining the discourse of adult educators.

The discourse of adult education

Many adult educators are committed to empowerment of their students through the educational process. So in adult education, if nowhere else, the silent murmuring should be heard. But as Chapter 2 on participation in adult education shows, this form of provision mainly attracts those who are already capable of making their voice heard. Chapter 1 shows how adult educators may be seduced by the hegemonic educational discourse to greater or lesser degree. Forces ranged on the side of conformity to the dominant discourse are formidable. A major national force is the use of education as a hegemonic instrument. Within this, the government-driven imperatives for a well-trained workforce and social conformity sit alongside the desire of powerful groups in society to preserve the *status quo.* The educational emphasis has changed to one which is now located on *individual* development (skills, competences, goals), self-help and student-centeredness, and a narrower definition of development.

Other forces are at work. The dominant discourse of adult educators is liberal humanism, based on student-centred learning, relevant to the individual, in supportive conditions and founded on a good relationship with the tutor. Various aspects of this discourse can cause problems. Some adults yearn for the 'good old days', when the teacher was the fount of all knowledge and through an authoritarian, didactic but pleasant approach, engendered feelings of security and confidence. It is not surprising that educators may be tempted by this attitude on the part of the student to retain ownership of knowledge and hence control of the discourse.

It is arguable that while student-centred pedagogy allows discipline to shift from the visible to the invisible and from overt to covert regulation, it does nothing to wrest power, knowledge and control of the discourse from the educator (Walkerdine 1988). Even the good relationship between educator and learner may be problematic. Mellin-Olsen (1990) advises students to relate to the knowledge communicated rather than to the teacher as an individual. He argues that if there is a relationship, then hidden messages from teacher to student inhibit the critical acceptance of knowledge by the student. Either 'I want you to understand this knowledge but in order to do so you must understand me at the same time' or 'if you reject this knowledge, you also reject me'.

Different discourses in mathematics

Discourses are sets of ideas, goals, values and techniques, competing ways of giving meaning to the world and of organising social institutions and processes. Hence when we are considering the context of cognition, we need to move beyond the wording of the problem to consider mathematics as a discourse. The discourse of mathematics is about more than just the rules of mathematics and the language in which it is expressed. Gilligan (1982) identifies fundamental different ways of reasoning. 'Separate' reasoning gets right to a solution in a structured algorithmic way, stripping away the context. It uses a mode of thinking that is abstract and formal, geared to arriving at an objectively right solution, dominated by an elaborate structure of rules and procedures and very judgmental. 'Connected' reasoning tries to experience the problem, relate it to a personal world, create context and remove ambiguity. It uses a mode of thinking that is contextual and narrative, geared to looking at the limitation of any particular solution and describing the conflict that remains, is tolerant in attitude toward rules, more willing to make exceptions and is reluctant to judge. Despite the best intentions of many mathematics teachers, the discourse of mathematics in the classroom and in text books frequently utilises separated reasoning whilst ignoring, rejecting or even despising connected reasoning.

Sadly this separated reasoning in mathematical discourse ensures that even after 11 or more years of mathematics education the discourse of school mathematics is unavailable to many. Harris (1995) found that when individuals were asked generally about skills used at work they often revealed a use of mathematics that they had previously denied when asked about it specifically in a mathematics questionnaire. The widespread negativity towards mathematics, supported by public acceptance that people cannot do it, allows it to be normal to answer questions about mathematics negatively, to deny any knowledge of the mathematics discourse. In addition, when mathematics caused the individual no problem, then the skills utilised were considered 'common sense'. Problems only became mathematics when the individual could not do them.

Induction to mathematics discourse

The underlying motivation driving adult students to learn mathematics in a formal situation is arguably to gain access to the powerful and prestigious discourse of academic mathematics (Northledge 1994). This approach helps to explain why they are less interested in building on their own existing everyday mathematical discourses which they feel, probably rightly, are not valued in society. So adults come to formal learning wishing to move themselves away from their 'folk' mathematics and, by gaining entry to the discourse of formal mathematics, operate in the world in new ways and gain social power and financial advantage through being able to speak as an insider to this discourse. Rewards for success in this endeavour come through the qualifications and certificates attached to the dominant discourse. However, learners experience difficulty in conquering the new discourse. Teaching needs to concentrate on conveying insights as to how the discourse works rather than content. If students are to become proficient in the discourse, they must learn to communicate it through speech and the written word. Learning then becomes the transition from informal everyday discourses to the abstract, systematic, rule-following rigorous discourse of academia. Learning starts with the discussion centred on the student and the discourses that they are familiar with but leads to the discourse of academic mathematics. So learners need pathways from their everyday discourses into the unfamiliar terrain of the academic discourse.

Teaching and the syllabus as a whole can be thought of as narratives which develop from the familiar, perhaps through case studies, and are plotted to lead students into the unfamiliar so that they may become users of the new discourse themselves. This approach to learning mathematics in which familiar practices become formal mathematics by a series of transformations is echoed by Walkerdine (1988). She argues that what is crucial is the metaphor which allows the task to be located within the framework of a familiar discursive practice. If used consciously, metaphors can illuminate and facilitate learning. They invite the use of intuition and imagination, developing new individual pathways from what is familiar and known to what is unknown. The use of metaphor allows the learner to position themselves with regard to the new knowledge. However, this approach can place some quite sophisticated linguistic demands on the adult learner in terms of communicative competence. If adults learn mathematics in part to acquire this communicative competence in mathematics language, then techniques are required to direct the learner's attention to the nature of the discourse while still retaining a normal level of communication. This requires the use of both commenting and meta-commenting by the tutor whereby the comments deal with the mathematics under discussion and the meta-comments draw attention to issues of the discourse itself.

Discourses are both enabling and constraining. The discourse of 'consumer education' common in many basic skills courses utilises the money economy, an area of life familiar to most of the learners. There is a clear social purpose to this section of the syllabus, namely to help consumers make better economic decisions. Here the discourse is that of the consumer society with value having economic or commercial connotations. Terms such as 'good value' and 'value for money' are located in the framework of economic principles. It would not usually, for example, define other notions of value, say in environmental terms. It does not offer a critique to consumption. To do this the tutor needs to move to an alternative discourse, say that of green politics (Lee 1995). The meta-comments can be used to raise the learners' awareness of the alternative discourses and facilitate their movement between these.

Mathematics as a language

The importance of contextualisation dominates the literature of mathematics education. There is uniform agreement that mathematics presented out of context is harder to learn, less transferable and boring. A quantitative illustration of the value of contextualising is illustrated by the following research. A group of adult students was asked to find the average of four numbers 15, 17, 18.25 and 15.75. Then they were asked to find the average weekly wage of four teenagers whose weekly wages were £15, £17, £18.25 and £15.75. Fifty-six percent of the adults got the first question correct but 70 percent got the contextualised problem correct (Rees and Barr 1984). Separating mathematics from its context leads to at best a mechanical facility with mathematics rules. If the context is sufficiently of interest to the adult learner, they will be motivated to apply a diverse range of analytical and mathematical skills to establishing a solution in this richer context. The most powerful learning tool might be that of the adult or group of adults generating their own written problems taken from their own current situations or preoccupations.

But the generation of genuinely contextualised problems is problematic for several reasons. 'Real' learner-generated problems need locating in wider frameworks and may not be neutral. After all, the problems that genuinely concern us are affected by who we are in terms of our sense of gender, class, age and ethnicity, and by individuals and groups that we have met and identified with. These and other factors will influence the choice of problem generated. The result may well involve large amounts of reading, writing and discussion to provide genuine contextualisation.

Such long questions for possibly 'only' a simple mathematics problem may be unacceptable in today's time-pressed teaching. To obviate this wordiness, oversimplified examples are often constructed for the learner by teacher or text which give the bare skeleton of a problem but do not reflect situations in work or real life where the mathematics is likely to be encountered. Before starting on any mathematical processes in these simplified problems, the learner must read, comprehend and extract the relevant information from the text. Application of usual reading aids such as context clues are severely limited, if not impossible. The written problem can allow for no ambiguity of interpretation and every word must be read, interpreted and its relevance to the problem assessed. In this sense mathematics is more like poetry than prose. Arguably this can make learning mathematics as difficult and unrewarding to some (and, of course, rewarding to others) as reading poetry. The lack of reading aids is particularly problematic for those with literacy difficulties.

Mathematics is not a first language

A commitment to relevant and contextualised mathematics is likely to result in problems embedded in wordy and perhaps culturally-specific problems. This raises linguistic and cultural issues. People from some minority ethnic groups have English as their mother tongue but for others it has to be learnt as a second or even third language. Naomi Sargant's extensive study of adult minority ethnic group participation in education (1993) found that many people, from many backgrounds, experience difficulties with literacy in their own first language, let alone in other languages. For minority ethnic group students or students studying in a second or subsequent language, this may be extended to the shift from experience of how academic discourse is organised in their culture to how it operates in the British culture. A national survey of Access students learning mathematics showed that for the Asian group, 82 percent had another language besides English as the main language at home (Benn and Burton forthcoming). So the level of English of adults learning mathematics may vary from little to fluency. In addition, some conceptual differences that adults have may stem from language. Linguists hypothesise that the structure of a person's first language has a significant influence on the cognitive process in mathematics learning such as classification and recognition of equivalence (Frankenstein 1989). Tensions between natural classification and those required by formal mathematics may lead to alienation or mystification.

Other linguistic issues

Even small changes in semantic structure in elementary addition and subtraction can lead to a considerable disparity in the number of correct answers.

> When a group of children were asked the following question 'Joe has 3 marbles; Tom has 5 marbles; how many marbles do they have altogether?' 97 percent gave the correct answers.
>
> The similar question 'Joe had some marbles; then he gave 5 marbles to Tom; now Joe has 3 marbles; how many marbles did Joe have in the beginning?' gained 83 percent correct answers.
>
> But 'Joe has 3 marbles; Tom has 5 more marbles than Joe; how many marbles does Tom have?' only gained 47 percent correct answers.
>
> (De Corte and Verschaffel 1991)

Problems may lie in the order of the words in the sentence. In the above problem, 'subject-verb-object' are the easiest to understand. The use of the passive tense and subordinate and comparative clauses raise linguistic rather than (or perhaps in addition to) mathematical problems. Hence, the ease of the problem is affected by the length of the problem, its grammatical complexity and the order of statements (Wareham 1993). On the other hand, De Corte and Verschaffel suggest that children spontaneously use a wide variety of informal methods to solve word problems and suggested that word problems should be introduced before formal operations and related number sequences (1991). Adults with their varied ethnomathematics could benefit from this approach. Certainly a wider understanding of the semantic structures of word problems and the strategies adopted by adults to solve them could help all adults learning mathematics.

A further linguistic issue is vocabulary. The learning and teaching of mathematics involves the use of both everyday English terms and more specialised mathematical terms. Kane (1970) introduced the notion of ordinary English and mathematical English to help analyse the special nature of written mathematics, arguing that they are sufficiently dissimilar to require different skills and knowledge on the part of readers to achieve appropriate levels of reading comprehension. But the problem lies not only in new words but also words that have different meanings in ordinary English and mathematical English such as product, mean and average. Other words, such as 'algebra', 'geometry' and 'bisect', have a mystifying and alienating impact. Time spent demystifying these words by a discussion of their roots may be time well spent.

Talking mathematics

We now return to the wider notion of discourse by reiterating the links between knowledge and power. Freire gives an example of peasants who call themselves ignorant and say that the 'professor' is the one who has knowledge and to whom they should listen.

> Why don't you explain the pictures first? That way it will take less time and won't give us a headache?
>
> (Memmi quoted in Freire 1985: 39)

Or as a student asked Marylin Frankenstein (1989)

> Why can't you give us the a, b, or c and then all we have to do is add, subtract, multiply or divide?

These comments illustrate that some learners see power and control as belonging to the tutor who, having previously digested mathematics knowledge, now regurgitates it in digested, easy-to-assimilate form to the students. This approach sees the teacher-student relationship as fundamentally narrative with the narrating done by the teacher and passive listening done by the students. This illustrates Freire's 'banking' concept of education where the teacher makes deposits and the students receive, file and store them (1972). He argues that the more students work at storing the deposits, the better students they become and they are rewarded with qualifications. But the more they work at this, the less they develop the critical consciousness that many educators see as the prime purpose of education. This may suit groups in society who benefit by the existing social structures and hence do not care to have the world revealed or see it transformed. A preferred alternative mode of education is problem-posing education based upon dialogue: teacher and student learn from each other and are jointly responsible for the education process. Banking education resists true dialogue; problem-posing education sees it as integral and indispensable.

Although adaptive rather than transformative, mathematics education in Britain has incorporated problem-solving enthusiastically into the curriculum. Nevertheless, at all levels and to all ages, the perception of the role of the teacher as the one who will eventually reveal the answer has not changed (Lerman 1992). However, the opening-up of the subject to discussion between students in itself is a move away from banking education. For adults, there is sometimes more freedom from the prescribed curriculum. Here the ideas of genuine problem-generation around issues of critical citizenship and common interest such as the finances underpinning the privatisation of British Rail or a Housing Association budget can be constructed by groups of students and solved in a critical and analytical way. The tutor can share his/her mathematical knowledge but is just one of the group in terms of the ethical issues and value judgements. Here the body of mathematics knowledge is subsidiary. It becomes a library of accumulated experience. When a problem is generated which reveals a need for some of this knowledge, then the context, relevance and meaning of mathematics knowledge is established (Lerman 1992).

Knowledge of the discourse of mathematics can be systematically developed by dialogue not only between teacher and student but also between students through group work. By working with others while learning mathematics, students can, through discussion and dialogue, break patterns of dependency, social relationships and isolated learning. Unlike the banking method, the use of dialogue develops a sense of ownership of knowledge and encourages questioning critical development. The use of groups also encourages problem-generation and -solving approaches to learning mathematics. If we see mathematics as a language, then learning any language would be no fun if it consisted solely of grammar and exercises. Similarly mathematics is no fun if it consists of just rules and exercises. The following metaphor or fable illustrates:

> Recently I attended a carpentry course. It was pretty tough.
>
> All the students (or almost all) were eager to learn. The first three weeks we learnt to drill holes. We found out about curious kinds of drills and bits, and how to make holes at odd angles. We got pretty good and accurate at drilling holes.
>
> The next six weeks were involved in cutting wood. We used all kinds of saws, found out how they interacted with different kinds of wood, and learned to cut accurately and smoothly. I got pretty good at planing wood.
>
> Joints was a different course. It took eight weeks, and we learned many kinds of joints. I was quite good at making joints.
>
> We did courses on other things too: sanding, turning, polishing, gluing and so on.
>
> Finally, we had an examination. We had to use some of these skills. I did reasonably well, fifth in the class.
>
> After the course ended, I went to see the Director. I told him that I quite liked the course in a way . . . but really I took the course because I wanted to make a table. He said only the top two or three in the course went on to do things like that . . . I said, 'What did we learn all that stuff for?' He said, 'Our course prepares students for making tables.'
>
> (Brown 1989)

Group work encourages the exploring and discovering of mathematics in a concrete and human process. Through discussion, adults can learn to articulate their points of view, listen to others, ask appropriate questions, recognise and respond to mathematically relevant challenges and in these ways develop their mathematical conceptions and their applications. However, this approach of group discussion is not always favoured by tutors. A recent national survey of Access courses found that group work was favoured by almost half the students, but a surprising 70 percent of mathematics tutors chose lectures/lessons as their first choice of teaching approaches (Benn and Burton 1993). This finding is similar to that of Desforges (1989), that though classroom discussion features prominently in the rhetoric of modern mathematics education, this form of learning is not actually utilised by many mathematics staff in schools. His research findings suggest reasons that could be applied equally to adult education.

A prime focus of much current adult education is the building and sustaining of an adult's confidence. A discussion necessarily involves the challenging of ideas and hence may clash with the tutor's higher priority of building confidence. A discussion also involves opportunities for the testing and sharing of ideas amongst participants with a broad range of experience. Tutors convinced of the necessity of starting 'where the student is at' may not wish some of the group to feel disadvantaged by their lack of knowledge or expertise. Lastly, to be effective, discussions do take time. A tutor with limited time and a fixed syllabus to cover may not allow sufficient time for real learning to take place or may, with the best intention, dominate and steer the discussion.

Writing mathematics

If the learner is to use mathematics effectively in everyday situations, they will need to be able to express mathematical ideas in a communicable written form. If they are to design their own mathematics problems, they need to be able to write mathematics. So the language of mathematics requires reading, talking and writing as well as mathematical skills. These need to be developed through practice. Problem generation by individuals or preferably small groups from areas of current interest to the learners can be formulated verbally then written out. If the problem is from real life, it is likely that the context-setting may be long. This encourages the learner to see mathematics problems as an integral, but sometimes small, part of a wider problem. Other students can then work through this problem, hence giving practice in reading and discussing mathematics. Other techniques can be utilised to encourage these skills. A technique which has been very successful in other areas of adult education is to encourage students, on a rotating basis and perhaps in pairs, to be responsible for taking class notes. These can then be given to the rest of the group either verbally or in written form and discussed at the beginning of the next session.

Keeping a personal journal of learning on a mathematics course can encourage the learner to examine not just the mathematics learned but also to reflect critically on their experiences of the course. By reflecting on their learning, the learner understands more clearly their mathematics anxiety, mathematical concepts and techniques and which learning styles suit them best. The journal can also help the learner to discover the solution to problems through the process of thinking about them in writing. The consequent organisation of ideas and notions about the topic or problem, though sometimes difficult, makes connections across the subject.

Adults' knowledge of mathematics is often fragmented and learners have a very confused mental map of the subject which may consist of half-remembered algorithms and little else (Coben and Thumpston 1994). The process of writing can make connections in this map but also show that there are often many valid ways to solve a problem. This can empower the learner to construct their own techniques rather than sieve through the map. It also provides a medium for the learner to assess their progress. Years of teaching mathematics has shown me that people's ability to underestimate their own abilities and progress cannot be overstated. This is encouraged by the prevalent attitude to mathematics that 'if I can do it, it can't be proper mathematics'.

A journal encourages the learner both to write in the discourse of mathematics but also ensures a concrete record of progress. It can also allow the learner to explore their feelings about mathematics. This can either be done in the ongoing learning journal or through the writing of a mathematics autobiography or life history. Through this safe and private medium, the learner can explore the role of mathematics and mathematics education in their lives, assessing the impact of others on it such as parental or teacher expectations, if their mathematics or lack of it has affected their life through, say, choice of further education or career and, just as valuably, any pleasurable memories of mathematics. Coben and Thumpsted (1995) found positive responses came mostly when people talked about their experiences of doing mathematics rather than learning mathematics but whichever, these positive feelings need to be recognised and valued by the learner.

Whose mathematics, whose education?

The role of the education system in constructing individual self identities cannot be underestimated. In an investigation of the mathematics histories of access students, time and time again students commented that they had been made to feel stupid by their mathematics teacher at school (Benn and Burton 1993). These comments were still remembered by these adults 20 or more years later and had caused the individual to construct a negative image of themselves with respect to mathematics (and perhaps more widely). Within our society, esteem or value is only given to those who grasp the correct forms of knowledge and ways of constructing that knowledge. These are rewarded with qualifications. Those who gain 'other' knowledge, ie ethnomathematics or folk mathematics, are not given qualifications. The ability to solve problems, if this is not expressed in the discourse of formal mathematics, is not valued.

One of the most significant features of the process of knowledge being owned, both in its practice and definition, by the academy, is the way that adults who have a wealth of experience see their own knowledge as somehow being different to 'real' knowledge which only institutional teaching gives them (Stuart 1995). It raises the fundamental question as to whether institutions of learning should define knowledge *per se* or be places to explore the context of knowledge with a range of competing voices. If the latter, then this allows the discourse of learning to be fluid not rigid. This approach allows issues of power relations and structural inequalities to be addressed in relation to the context of knowledge acquisition.

The autobiography can be a useful tool in this approach, allowing the learner a way of contextualising their learning and experience of mathematics within a social framework. By telling their own story, learners can explore the construction of their mathematics knowledge and how experience has shaped this. This then allows the learner to become more aware of and more responsible for their own learning. Hence the autobiography seems ideally suited to revealing experiential learning and tracking the development of the self as learner.

There is a caveat, however. If the autobiography is seen as a self progressing through life with a clear set of aims and ambitions, then again this concept of linear progression may reflect white male experience. Many women and members of minority ethnic groups may find it more appropriate to express their autobiographies as a collage or montage rather than linear progression which represents the dominant narrative (Usher 1995). This echoes a point already made earlier that rather than allowing a single form of narrative which normalises all narratives, if the autobiography is to be an effective learning tool for all, then alternative constructions offer the potential to illuminate the complexity of most adults' mathematics lives.

The use of non-linear visual metaphors such as the quilt enrich our understanding and offers ways of seeing mathematics life history in a non-linear way (Sellars 1995). It allows chronology to be overridden and influential events to be highlighted. It also allows the interaction of the rest of life with the mathematics life history. Autobiography provides a way of reflecting on the past. In mathematical autobiographical work, attention is focused on one's personal mathematics history and the forces and contexts that have patterned the way we see and do mathematics.

But the way we see and do mathematics is influenced by who we are, our social, economic and political location in society and our cultural background. The complex interaction of these factors with learning mathematics will be explored in the next chapter.

Chapter Eleven

Mathematics divides society

Culture

The previous chapter argued that the way we see and do mathematics is influenced by who we are, our social, economic and political location in society and our cultural background. The next four chapters will examine the effects of cultural difference on adults learning mathematics. Differences based on class, ethnicity and gender are important and will be dealt with later but this chapter will set the scene within the wider context of a general approach to the examination and exploration of culture, mathematics and adult education. We will investigate the reasons for the dominance of academic Western mathematics, identify other mathematics or ethnomathematics and suggest ways that an adult education curriculum can build on the strengths of both.

Britain today could not be described as monocultural. There are a wide variety of immigrant communities including African-Caribbean, Arab, Asian, Chinese, Greek, Hungarian, Irish, Italian, Polish and Turkish. There are four countries, different languages with regional and cultural variations, dialects, religions and traditions. Society has increased in cultural diversity and the resultant cultural differences have come to manifest themselves more strongly in education. One of the dominant questions in all education provision is how to accommodate difference to enhance education for all. The pressure has mounted for education to reflect the multicultural nature of society and to re-evaluate educational experience in the face of the educational difficulties of many from minority ethnic or working class communities. There is a growing sense that all education, including mathematics, needs to be studied in the living contexts which are relevant and meaningful to the learner, no matter what their cultural background.

This is not just a preoccupation of educators. A mature and humane democracy depends for its survival upon its citizens. Investment in education for pluralism and diversity is critical if citizens are to feel a sense of togetherness rather than alienation. A society which has neglected this dimension of education may begin to lose its cohesion. It can be argued that cultural education is an effective social glue (Avari 1995). This is particularly true for adult education. Adults need to reflect upon their identity, to show respect for the identity of others and to learn to find common ground on which their private and public identities can rest in harmony (Modood 1992). A multicultural curriculum for adults in Britain needs to have at its core an appreciation of the historical, political and economic dimensions of the multicultural society and its links with the global society.

Mathematics is a kind of cultural knowledge which all cultures generate but which need not necessarily look the same from one cultural group to the other. As all cultural groups generate language, belief and rituals so they develop mathematics. Hence mathematics is a pan-human phenomenon. Just as the language generated is that of the group, so is the mathematics. However, not all mathematics are valued. Intellectual racism, sexism and classism in educational institutions have affected the way that knowledge has been perceived, produced and propagated by scholars and academics. Eurocentrism locates the source and subsequent development of all worthwhile knowledge in the historical and geographical space of Europe and its cultural dependencies and both consciously and unconsciously denigrates the knowledge of 'others' (Anderson 1990; Avari 1995; Joseph 1987, 1990, 1991). Similarly our patriarchal and hierarchical society has normalised male middle-class knowledge and values and pathologised or ignored that of women and the working class. Different cultural backgrounds lead to different languages, experiences, behaviours, attitudes, values and skills. Stereotypes of able and less able are strongly related to the learner's anticipated ability to achieve particular examinations and learn in particular ways. These preconceptions are in turn related to a variety of cultural assumptions linked to class, race and gender stereotypes (Scott-Hodgetts 1992).

Recent advances in theories of cognition show how strongly culture and cognition are related. The reductionist tendencies of modernity have tended to dominate education and mathematics and until recently have implied culture-free education and culture-free mathematics (D'Ambrosio 1990, 1991). In the last decade or so, alongside the still-dominant scholarly, Western, academic mathematics, the existence of ethnomathematics, practised by identifiable cultural groups, focused on interest, motivation, codes, jargon and discourse which is not that of academic mathematics, has been acknowledged.

The development of academic mathematics has links with some ethnomathematics. New ideas can appear outside academic mathematics, be used and practised as ethnomathematics and eventually be recognised, theorised and expropriated as academic mathematics. However, these ideas sometimes do not become formalised. The practice continues whilst useful, but restricted to the cultural group which originated it. This illustrates mathematics as a social construct rather than a structured body of knowledge. It is a set of *ad hoc* practices which evolve as a result of societal change (D'Ambrosio 1991).

Mathematics education, like all education, is the process of inducting the young and not-so-young into part of their culture. What we have termed academic mathematics is perhaps more appropriately termed Western middle class male mathematics. Other cultures may see this as alien to a greater or lesser degree. This raises issues of enculturalisation (induction into the home or local culture) and acculturalisation (induction into a culture different from the home or local one).

Culture and mathematics

If mathematics is seen from a formalist perspective as absolute and given, then it is very much received knowledge. Inability to do mathematics can be linked to innate hierarchical abilities and the abstruse nature of the subject itself. If, however, the view taken is that mathematics is founded in human activity, mathematics is no longer a special case, a peek into the mind of God, but is liable to growth and change within the social context. The social nature of mathematics allows the cultural differences of individuals to be given space and hence enrich the mathematics curriculum. Bloor (1976) argues that 'the world does not, for the most part, consist of isolated cultures which develop autonomous moral and cognitive styles. There is cultural contact and diffusion. In as far as the world is socially blended then to that extent it will be cognitively and morally blended too'. Hence recognising the likelihood of commonality in mathematics thought, he identified five differences which have arisen through social causes:

- variation in the broad cognitive styles of mathematics;
- variations in the framework of association, relations, uses, anomalies, and the metaphysical implications attributed to mathematics;
- variations in the meanings attached to computations and symbolic manipulations;
- variations in rigour and the type of reasoning which is held to prove a conclusion;
- variation in the content and use of those basic operations of thought which are held to self-evident logical truths.

These could identify particular factors that could act as guidelines for catering for the cultural individuality of learners.

This thinking is echoed by postmodern philosophers such as Foucault (1986) and Lyotard (1984) who argue that the divisions of knowledge disciplines accepted today are modern constructs defined from certain social discourses and that all human knowledge consists of narratives each with its own legitimation criteria. Mathematics is generated by human activity and is interconnected to all human knowledge. It is hence an outcome of humans and their cultures and hence culture-bound. Mathematics is constructed, as all disciplines are, to help the culture make sense of life and the world, and to provide tools for dealing with the full range of human experiences. The purposes may be religious, artistic, practical, technological or study for its own sake. Whatever, the mathematics of each culture is based on the values, goals and purposes of that culture. Western academic mathematics is very powerful in our global society but its value can only be seen relative to its own culture. It may be literally useless in another culture. A corollary here is the English language which is very powerful globally but not intrinsically more valuable or more useful in all contexts than other languages (Ernest 1991).

Adult education, culture and mathematics

Cultural issues have always been of prime concern to adult educators. The idea that disadvantage is collectively experienced by different groups in society is well expressed in the following quote which is directed at access to education generally.

> For the problem of access is seen primarily in terms of the access of individuals (either as such or else as members of particular groups) to learning opportunities, the barriers to which are constituted in wholly material terms. In fact barriers to access are collectively as much individually experienced and culturally as well as materially constructed, and the paradox lies in confronting the individual learner with the problem of the socially and culturally constructed contents of learning. All the barriers likely to be experienced in these circumstances are equally real, but access does seem to be conceptualised in terms which ignore the cultural barriers, which isolate and abstract the individual learner, and which tend to reduce the issue to one wholly resolvable in technical and institutional terms. There is little sense here that access to education might be a collective and political issue of knowledge and power in society.
>
> (Griffin 1983)

Adult educators struggle to fight continuing marginalisation caused by racism, ethnocentrism and sexism and understand how the knowledge of certain groups together and separately are defined as 'other' practices and hence systematically devalued. To promote equity, adult educators need to understand how marginalisation is discursively constituted through professional and personal taken-for-granted beliefs and actions. The liberal pluralism and the ideas of student-centred learning that have been the foundation of adult education mask how provision inescapably privileges the learning of some at the expense of others.

It is not sufficient to accommodate the existence of other cultures by providing different cultural contexts for problems or illustrating multicultural roots throughout the history of mathematics. It may be that cultural differences in cognition reside more in the situations to which particular cognitive processes are applied than in the existence of a process in one cultural group and its absence in an other. Some examples illustrate:

- In an exchange between a native African Demara sheep herder and a European, the herder agrees to accept two sticks of tobacco for one sheep but becomes confused and upset when given four sticks after a second sheep is selected. This is often quoted to show the herder's lack of understanding of arithmetic but could equally show the European's lack of understanding of barter where sheep are not a standard unit and hence not always valued at two sticks of tobacco.

(Frankenstein 1989).

- A European asks a Kpelle man this question to test deductive reasoning. All Kpelle men are rice farmers. Mr Smith is not a rice farmer. Is he a Kpelle man?

The Kpelle farmer's reply included the following: 'if you know a person, if a question comes up about him you are able to answer. But if you do not know a person, if a question comes up about him, it's hard for you to answer.'

The interpretation of this answer could reflect negatively on the Kpelle's ability to reason or illustrate different views on talking about people you do not know.

(Frankenstein 1989)

- When given an ethical question, boys use separated reasoning of the either/or variety whereas girls ignore this and try to work out alternative strategies which allow for conflict, stress and compromise.

(Gilligan 1982)

- In a court case about land rights, an Inuit hunter was unable to say how many rivers were in the disputed area, which was taken by the opposition as clear evidence that the man was unfamiliar with the region. In fact, because he probably knew the actuality of each river, there was no use to him in knowing the number of them. We count things when we are ignorant of their individual identity.

(Denny 1986)

- 'in various . . . parts of Australia, the natives show habitual uncertainty as to the number of fingers they have on a single hand' (Smith 1923 quoted in Harris 1987). This ignores the fact that in many Aboriginal languages the word for five is 'hand'. Asked how many fingers they have, apart from finding it a stupid question, the answer would be 'hand'.

(Harris 1987)

In all these examples, the 'others' are judged to have failed but by the standards of just one group in society. They are victims of the assumption that logic and mathematics have no social preferences. It just so happens that certain sectors of the population (whites, males and the middle classes) are intrinsically better equipped for these forms of study. Their cognitive styles embody the properties described as mathematical or logical styles. The fact that this cause and effect may actually be reversed (ie, mathematical or logical thinking has been defined as that of male white middle class cognition) is usually ignored (Ernest 1991).

Mathematics education: the need for a change

The strongest case against current mathematics education is that it is useless and meaningless to the majority of students. Fasheh (1991), developing Freire's work, argues for education based on praxis, 'the combination of concrete conditions (social, cultural and material), reflection and action in constant interplay'. The goal of education is to make sense of the world, our experience and our culture and to respond to the challenges posed by the real environment. Consequently, it is necessary for the learner to critically appropriate knowledge outside of their experience in order to broaden understanding of themselves, the world and the possibilities for transforming taken-for-granted assumptions. Individuals need to value and have valued their own ethnomathematics. Then after acquiring academic mathematics would choose for themselves the most appropriate to use in different circumstances.

The notion of mathematics as a dynamic which arises out of human activity acknowledges the existence and validity of different mathematics generated in different contexts and cultures and hence lays the framework for the acceptance of ethnomathematics. This approach forces a re-examination of notions of human abilities. In contrast with the positivist approach, the constructivist one assumes that mathematics abilities are multi-dimensional and changeable. Instead of focusing on individualism, it encourages co-operative learning and the social construction of knowledge. This use of mathematics develops an understanding of the world and hence an awareness of inequalities in our society and the underlying assumptions of social organisation which cause them. This may lead to the creation of new ideas, perspectives, insights, images and models (Volmink 1990). It exposes the ideological dimension of mathematics and the relationship between knowledge and power and recognises that hegemony is not only characterised by what it includes but what it excludes, by what it renders marginal, deems inferior and makes invisible. Mathematics can be used to help develop a wider multi-cultural perspective and enable students to see how powerful the subject can be as a tool for examining society. If this approach to multi-cultural mathematics is adopted, then a teaching and learning for mathematical, educational and democratic purposes may emerge which is more than tokenistic.

Adult education is based on interaction and depends on the already-acquired coping mechanisms brought into education by every adult. To ignore this individual history is not only unproductive, it is to ignore the factors which have built up the individual's personality and basic knowledge. Without this, the images and metaphors for education are unavailable. Adults will not easily or willingly change their modes of thought which they have developed in their culture for explaining and coping with their realities, to accept approaches constructed to deal with alien realities of other cultures.

Eurocentric mathematics

Western mathematics has been one of the most powerful weapons in the imposition of Western culture. Up to 15 to 20 years ago, mathematics was thought to be culture-free and with universal validity. Colonial powers considered it appropriate to teach the same (for 'same', read 'Western academic') mathematics anywhere in the world. But this mathematics only seems universal because it consists of abstractions from the real world. But whilst abstractions may appear so, they are not of necessity context-free and universal.

Anthropological literature demonstrates that academic mathematics is not the only mathematics (Bishop 1990). Many counting systems exist in the world. There are other conceptions of space and logics. That these alternative symbolisations of arithmetic, geometrics and logics exist, implies at least the possibility of alternative mathematics. Cultures have generated mathematical ideas as they have generated language, religion, morals, and values. But the mathematical ideas of other cultures have been influenced by the more dominant European notions. This is particularly true of ex-colonial cultures.

The three major mediating agents in the process of cultural invasion in colonised countries by Western mathematics were trade, administration and education (Bishop 1990). Trade was almost all carried out in Western measurement systems and currency, which would have had a major impact on local mathematics systems. The mechanisms of administration and government (keeping track of large numbers of people and commodities) would have necessitated Western numerical procedures in most cases and hence the adoption of the Western number systems. In addition, education itself was a major medium for cultural invasion. In post-war Africa, for example, it was suggested, even at the time, that all Western education could be seen as an instrument of violent social change splitting the individual from his (not usually her) local culture and those that had received it from those that had not (Benn and Fieldhouse 1995). The use of English introduced a linguistic apartheid between the elite and the rest. Mass education was introduced and encouraged by the Colonial Administration with the prime intention of enabling local people to function adequately in European-dominated trade, commercial and administrative structures which had been established. Any mathematics taught was with the intention of improving the quality of the African workforce and thereby the economy and was limited to arithmetic and its applications. The secondary and higher education given to very few was aimed more specifically to provide an educated elite for the government service.

The mathematics curriculum was, at its worst, elitist, irrelevant and abstract. This was exacerbated by the absence of suitable text books with an emphasis on African rather than British concerns. Questions in Tanzanian colonial textbooks were, for example, based on cricket, escalators at Holborn tube station and Imperial units of measure! (Bishop 1990) The situation in colonial Africa was echoed throughout the world and under all colonial powers. Colonial education was a process of acculturisation rather than enculturisation. It conveyed the Western value system with its emphasis on rationality, deductive reasoning, logic, objectism, decontextualisation, power, control, progress, change and precision. From colonial time through to today, these attributes of mathematics are so powerful that this academic or Western mathematics is taught everywhere and is still assumed by many to be universal and culturally neutral.

There was a further effect of colonialism on academic mathematics. Most histories of mathematics which were to become standards for later works were written in the late nineteenth or early twentieth century. This period saw the culmination of European domination of both Africa and Asia by the colonialist powers and the growing ideology of racialism, white superiority and cultural domination. Hence the histories of mathematics emphasised the unique role of Europe in the development of mathematics whilst ignoring the contributions of the colonised.

The lessons of colonial invasion of the cultures of the world are of value today in any examination of multicultural education. People in a position of power perceive only a limited reality. This creates unbalanced knowledge. The world looks different for those who are members of less influential groups. On their horizons loom inescapable boundaries and those who impose the boundaries. But these less powerful groups have their own multi-layered and complex worlds. They have their own knowledge. They have their own mathematics. The power locations of knowledge generation need to be recognised, relocated and redistributed (Wolffensperger 1993).

'Other' mathematics

The importance of cultural context is fundamental to pedagogical practice. If ethnomathematics can be identified and located in their historical origins and cultural discourse, then this may allow for the redefinition of legitimate mathematics knowledge and practices. Writers and researchers such as Gerdes (1986) who looked at the indigenous 'frozen' mathematics of Mozambican weaving, Saxe (1988) and Carraher's (1991) work in Brazil, Scribner's work (1984) in factories and Lave's (1988) work with supermarket shoppers all suggest that many seemingly innumerate people actually do engage in complex mathematical thinking but do not use the conventions or techniques of academic mathematics to articulate or solve their problems. It maybe that it is the format of academic mathematics rather than the underlying concepts that many people do not engage with. The use and legitimisation of ethnomathematics in formal learning would not only aid mathematics learning and understanding but also free learners from the tyranny of conventions and formats of academic mathematics.

But in themselves, ethnomathematics are not sufficient. Adults wishing to learn mathematics have diverse motivations and goals and, of these, problem-solving may be only a small part. Academic mathematics inducts the learner into the game of academic mathematics with its own rules and rewards. As these rewards include the acquisition of qualification for further education or career advancement, the overall goal of the learner may be to learn to play the game rather than to learn mathematics *per se.* In this case, concentration on ethnomathematics to the exclusion of academic mathematics would be to the detriment of the learner.

Academic mathematics is a socially organised mathematics defining the social institution of mathematics (Abraham and Bibby 1992). Though usefulness and effectiveness may lie with ethnomathematics, power and status lie with academic mathematics. A curriculum focussed totally on ethnomathematics would almost certainly be ghettoised. Many adults wish to make a greater sense of their own lives and experiences. The valuing of ethnomathematics would facilitate this but for a firmer understanding, learners need to understand how and why the other prevalent mathematics have held such dominant sway. They need to be able to understand, operate within and, if they wish, criticise the social institution of mathematics. This ability would allow individuals to criticise the social structure of society and to participate in active democracy. Learners require a command of both ethnomathematics and academic mathematics. They also must have access to a value system which allows critical judgements to be made as to which to use in different situations.

Examples of ethnomathematics

One criterion of academic mathematics is that it is written, not oral. However, various interesting findings emerged from an investigation of problem-solving strategies in out-of-school mathematics in Brazil (Carraher 1991). Interviews with young street vendors showed that the vendors solved correctly 98 percent of a set of problems given in the street setting, 74 percent of verbal problems given in formal settings but with explicit reference to real world objects and quantities and only 34 percent of the straight computational questions. Follow-up studies showed that oral techniques were used more frequently than written ones and were more successful, even when used by youngsters not in employment of any sort. When the children chose to solve the problems in their heads rather than by written algorithms that they had been taught in school, they were more likely to be successful. It was found that children used written and school-type procedures in situations associated with school (the solution of computational problems) and extra-school methods in extra-school situations (shop sales). Carraher concludes, however, that there are fundamental similarities between the oral ethnomathematics and written academic mathematics.

Children who could do mathematics outside school, though showing a great deal of difficulty with school approaches, did demonstrate a knowledge of the fundamental invariants of addition and subtraction. The difficulties they showed when attempting school mathematics related to the deployment and meaning of particular procedures and representations adopted by schools rather than the fundamental arithmetic of addition and subtraction.

Carraher's (1991) and Saxe's (1988) work in Brazil seems to shows that different jobs place varied demands on workers. Even those with little school mathematics develop an appropriate ethnomathematics to deal with these demands. The major advantage of this mathematics is its meaningfulness, its major liability the limited conditions to which this knowledge may be useful or relevant. Identifiable subgroups or cultures have built up theories-from-practice which can be passed on to others but may not be generalisable. Academic mathematics, on the other hand, is based on theory-to-practice but this also has limitations. For example, students are frequently required to use symbolic representation before they have developed a full understanding of their potential meaning. This may lead to procedural rather than conceptual knowledge.

Scribner (1984) echoed these findings in her studies of factory workers. She found that the ethnomathematics of the 'unskilled' workers was always more effective than the use of academic mathematics by 'academically-skilled' individuals who were not part of the group who normally perform the task. The 'unskilled' workers had developed effective approaches to the solution of mathematical problems involved in their job and had adapted these as the task evolved. The more 'academically-skilled' were more likely to be single algorithm problem-solvers. They had been better-trained in academic mathematics which values the use of the single algorithm as a property of generalisability but this did not give them an advantage in the practical situation.

Lave (1988), working with adults whom she judged were expert grocery shoppers, found that arithmetic problem-solving in formal tests and shopping situations was quite different. Their score averaged 59 percent on the arithmetic test compared to 98 percent for the arithmetic in the supermarket. The form and content of the arithmetic used in the shopping problem had grown out of the situation. The shoppers, again working in their heads, had developed iterative problem-solving methods. These, by gradually closing the gap between the problem and the anticipated form of solution under continual monitoring, accounted for the high number of correct responses. This process of iteration and monitoring, whilst gradually moving towards the solution, can be seen in the dialogues from Carraher's Brazilian work (1991).

Different cultures are constituted by different practices, meanings and discourses, and individuals are positioned both structurally and individually by their cultures. The transfer between different mathematical discourses is very complex. Adults wishing to learn academic mathematics are wishing to transfer between discourses. In any learning group, there will be individuals from different cultural backgrounds of gender, ethnicity, age or class. The discourse of academic mathematics may not, or perhaps cannot, be the same as that of work mathematics or home mathematics. Teaching adults mathematics as a process of transfer from each individual's ethnomathematics to academic mathematics whilst still valuing and legitimising the ethnomathematics is clearly no easy task. Educational development does not mean a linear extension of lived experiences. it means empowering adult learners with new experiences which enrich by contradicting or supporting earlier ones. Educational design needs to articulate with learners' earlier ways of performing in reality to optimise learning.

Usually, ethnomathematics are adequate and can be valued in their own right as appropriate and effective in the local culture. An example of this is the ethnomathematics of building workers in Latin America who construct octagonal corners or *ochava* (Llorente 1996). Interviews with workers with very limited school education showed that to construct a corner they used a basic formula using the fact that if the measurements 6 and 8 were maintained for two sides of the square, the third side of the triangle would always be 10. The workers could apply the principle of proportionality to show that the result would be the same for 3–4–5, 6–8–10 or 12–16–20. The research indicates that this is not a mere mechanical application of rules but rather an organising activity for correctly using the information provided by the particular social setting. The consequences of not applying the rule and hence determining the square could also be thought through: the building would be unstable.

In this example, there is no advantage to the workers in changing the system or introducing a layer of 'Pythagorean' theorising. This is not true for the members of the Landless People's Movement in Brazil (Knijnik 1996). Here the lack of academic mathematics has serious social and political consequences. In the words of one of the members of this movement:

> Well, my friends, in the research we had done in the settlements and campings where we were, we could observe the deficiencies among our comrades. Then, we realised that what our settlement companions really need is mathematics. They also need writing and reading, but mainly mathematics. They look for mathematics the same way they look for a medicine for a hurt because they know where the hole of the projectile is, by which side they are exploited.

An example of this lies in the *cubacao da terra*, estimating the area of a piece of land.

Figure 4 gives two different local methods (those of Adao and Jorge) which give different results both from each other and from the result gained by using the academic mathematics formula (*dos livros*). The results using the ethnomathematics are approximations which become more accurate the closer the shape is to a rectangle. Knijnic argues that it is not good enough to value these other ethnomathematics as a means of ensuring the survival of cultural traditions. This can lead to ghettoisation and disempowerment. There are economic and social disadvantages produced by the continuation of these

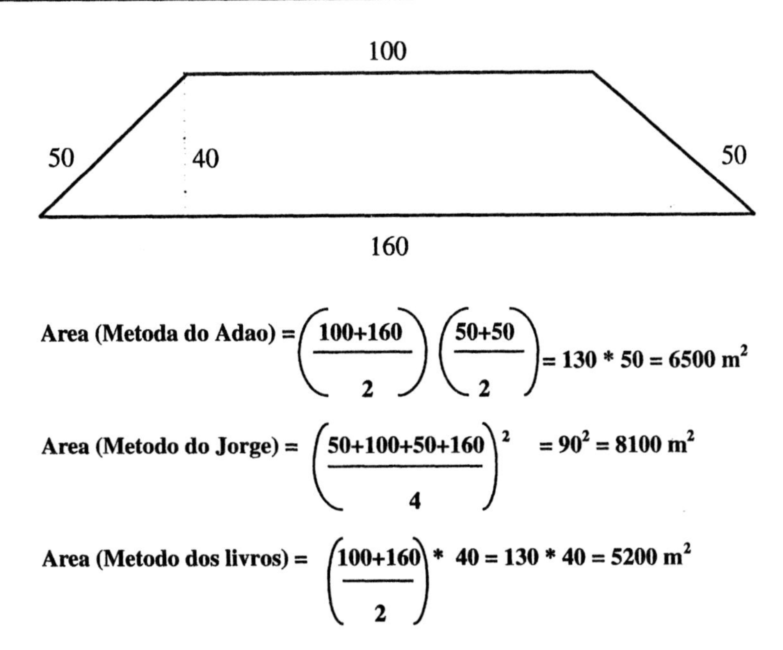

Figure 4. Alternative methods of estimating the area of a piece of land.

practices as compared with official land estimation methods. Therefore, these different knowledges produce and reproduce different power structures. She teaches her students to interpret and decode the ethnomathematics of their group and ensures that they understand that these lead to approximations. This facilitates movement between the different mathematics. This places the emphasis on valuing and incorporating into formal learning, the cultural aspects of the learners and building a bridge between the local and the broader knowledges. The success of this approach is shown by the fact that peasants walk for up to two days to attend her classes and is reflected in the following quote from one of her peasant students:

> . . . there is part of our life, part of your life, of the more elaborate knowledge you already had . . . You left your story to hear our people's stories. And now, your stories already have a bit of a mix of our stories . . .

This example exposes that mathematics are not just different, but, in terms of power, unequally different.

Mathematics education: valuing the practical

There is a critical dilemma between relevance and rigour. All educational literature argues for the importance and centrality of relevance in the mathematics curriculum. Relevance is seen as making content relate to the world of practice, helping the individual develop appropriate skills and capabilities and enhancing effectiveness in coping with life. Rigour, on the other hand, can be characterised as making content relate to the world of formal theory (Usher 1989). This results from the assumption in our society that theoretical knowledge (here, academic Western mathematics) is fundamental and can be applied to the instrumental problems of practice. Theory is hence 'real' knowledge and practice the mere application of this knowledge to the solving of problems. The implication is that there is no knowledge contained in practice. The result is the privileging of theory in order to give rigour to content. Rigour is the main priority and learners must somehow relate the theory to their practice rather than develop theory from practice. This one-way construction makes it difficult for learners to relate theory to practice or *vice versa* either during or after their studies.

Usher (1989) critisises the theory-into-practice model by arguing that theory cannot be mapped directly onto practice. In practical situations, facts are often incomplete and the situation often not fully understood and, in any case, theory does not tell us *how* to practise. Practice, in reality, consists not of the application of theory but of such things as attending and being sensitive to the situation, or anticipating and making *ad hoc* decisions. Hence academic mathematics, rather than being a reflection of reality, is actually a ritualistic language with little resonance in practice. It is part of a paradigm which is limited in its usefulness. Within this paradigm, ethnomathematics is seen as specific, intuitive and unsystematic and hence not true knowledge.

However, many adults need to use mathematics in order to act appropriately in the world. Hence ethnomathematics is knowledge not of the world but of how to act in the world. Because it is concerned with appropriate or 'right' action, it must have an ethical dimension. Because all people are 'situated', this dimension must depend on the situation and hence value decisions are unavoidable. The mathematics used cannot be universal but must always be mediated in the light of a particular situation. When individuals use mathematics in real life, the knowledge they use may involve academic mathematics but it is always mediated in the light of the circumstances of the situation and is, therefore, situational, value-laden and hence ethical. Understanding and using mathematics always involves interpretation and application and this takes place within one's own perspective, framework or paradigm which is influenced by one's culture. Hence ethnomathematics are not necessarily merely routinised, habitual activity but realms of knowledge in their own right with their own appropriate modes of reasoning and understanding and own value systems.

Ethnomathematics do have potential limitations. They may not encourage critical reflection, may limit awareness and be limited in scope and depth. They may be either committed and informed action or habitual and routine practice. Nevertheless, starting with practice problems rather than practice *per se*, frameworks of practice can be analysed. Denormalising both ethnomathematics and academic mathematics could lead to a critical approach to both forms and to the beginning of a dialogue which could suggest alternative possibilities. It could also lead to an introduction into the discourse of academic mathematics which, whilst it cannot be applied directly to or mapped directly onto the ethnomathematics, may illuminate it. Understanding the framework of academic mathematics, why it takes the form it does, and how it is as much a construct and therefore, as much capable of problematisation as their own ethnomathematics, could enable learners to recognise that the two mathematics are telling different stories but one no more right than the other. The stories have different sites of origination but the same site of application (Usher 1989). By the process of placing academic mathematics into a context through explaining its assumptions, concepts, values and language, the ethnomathematics can be located within or alongside this framework. Both can be examined not as a source of ultimate truths but in terms of their utility in developing new knowledge, understandings and changes in practice.

An illuminating example of simple arithmetic processes with different cultural resonance is given in this conversation with a university student in Papua New Guinea:

> I asked him how he would find the area of a rectangular piece if paper. He replied:
>
> 'Multiply the length by the width.'
>
> You have gardens in your village. How do your people judge the area of their gardens?'
>
> 'By adding the length and width.'
>
> 'Is that difficult to understand?'
>
> 'No, at home I add, at school I multiply.'
>
> 'But they both refer to area.'
>
> 'Yes, but one is about the area of a piece of paper and the other is about a garden.'
>
> So I drew two (rectangular) gardens on the paper, one bigger than the other.
>
> 'If these two were gardens which would you rather have?'
>
> 'It depends on many things, I cannot say. The soil, the shade . . .'
>
> 'I was then about to ask the next question, 'Yes, but if they had the same soil, shade . . . ', when I realised how silly that would sound in that context. (Bishop 1985)

The cognitive compartmentalisation of the comment 'At home I add, at school I multiply' can be a strength if recognised and valued but a weakness otherwise.

Cancelling the dividing factors

This chapter has identified the cultural base of knowledge and the consequent need for combining both academic mathematics and ethnomathematics in an adult curriculum. Institutions and tutors need to utilise and value the cultures of the learners. This involves understanding world views which are appropriate to the individuals themselves. Where this is not the world view of the tutor, it will require extensive understanding on the part of the tutor through appropriate reading, television, film or other means and by the use of student-generated problems and learning. This approach would encourage dialogue and experimentation, recognising and valuing skills which learners already have as well as those which tutors wish to develop. The range of bilingual, non-Eurocentric, gender and class non-specific materials and texts and resources would need to be extended. It would be necessary to acknowledge the values inherent in academic mathematics, its cultural history and develop open discussions about the relationship of society, educational institutions, tutors and the learners with these values.

To continue with the assumption that mathematics is value-free is to give implicit support to the values that we suggested earlier pervade mathematics. Cultural hegemony is the more powerful if unacknowledged (Bishop 1990). An extension of this would be the acknowledgement that examples are often not politically neutral (Maxwell 1988). This could be counteracted by the critical use of statistics to draw inferences and trends in issues of real social and political importance. The history of mathematics and the cultural assumptions of many mathematics historians could be examined. This could be illuminated by an exploration of the difference in language and structure of counting systems found across the world, eg, different calendars and different spatial relations in, say, traditional African designs, Indian rangoli patterns and Islamic art. A demythologising of the 'Greek miracle' could give rise to a more balanced assessment of early mathematics accomplishment (see Joseph 1987, 1990, 1991).

Dorn contended (1991) that discrimination has been made invisible by a liberal ethos that makes it difficult even to discuss the possibility of unequal treatment. The next three chapters explore issues relating to major social groupings: gender, class and race. Chapter 14 on race is rooted in a discussion of racism in our society and the need to combat this. All the arguments of this chapter apply but set in the context of structural racism. Similar arguments apply to gender, the subject of the next chapter. But important though consideration of race and gender are, it is arguable that the underpinning cultural divide is that of socio-economic factors ie, class. In our 'classless' society of the past Conservative era, the notion of class has almost disappeared from the agenda, but Chapter 13 will argue that it is still a way of expressing one of the fundamental divides of our society.

Chapter Twelve

Taking women away from mathematics

The relationship between gender and mathematics education, though extremely complex, is nevertheless firmly located in the socio-political context. In the past there have been arguments about the inclusion for mathematics in the curriculum for women. For example, in 1868 Dorothea Beale, a leader in the higher education for women movement, stated

> . . . I do not think that the mathematical powers of women enable them generally (their physical strength, I dare say, has a great deal to do with it) to go so far in the higher branches, and I think we should be straining the mind (which is the thing of all things to be most deprecated) if we were to force them to take up such examinations . . .
>
> (quoted in Willis 1989: 2)

Whilst this opinion is now thankfully no longer held, there is still a common assumption that the teaching offered to men and women, boys and girls, is experienced in the same circumstances. This view assumes that equivalence of outcome is assured if difference is ignored. It is typified by offering the same resources to all, comments such as 'I concentrate on the student and ignore their sex (or race, etc)' and declaring an equal opportunities intent without considering the implications of this on the institution, staff and students.

There are deeper issues involved in learning outcomes of students as opposed to provision by teachers and institutions. Learning opportunities are orientated to the dominant values in society. These are reflected in the education system as a whole and are concerned with helping individuals to adapt to dominant educational and cultural structures without questioning the modes by which education controls differential access to knowledge and power. This ensures that the rhetoric of 'equal opportunities' and 'individualism' obscures the way in which terms like 'individual need' and 'student centred' are socially constructed and located in ways that make mathematics more readily available to certain groups (Keddie 1981).

The constructivist approach to teaching argues that it is impossible for any two learners to construe the same learning experience in the same way. In particular, the different messages carried for males and females by social, political and economic contexts are part of the circumstances in which learning is set and hence are strongly influential on the outcomes. Learner expectations, societal expectations and tutor and institutional behaviour all fit in and are influenced by this context. The formal learning setting is a socio-political setting within which different messages are given to and received by different groups of students (Burton 1990). As a report on gender and schooling concludes, 'Girls, may follow the same curriculum as boys – may sit side by side with boys in classes taught by the same teacher – and yet emerge from school with the implicit understanding that the world is a man's world, in which women can and should take second place' (Stanworth 1983: 58).

In the 1960s it was generally believed that girls were mathematically less able than boys and research concentrated on 'why girls can't do mathematics?'. By the 1970s, biological determinism had been largely discredited and supported by the growing belief in social explanation for gender differences and the re-emergence of the women's movement. Researchers concentrated on achievement rather than ability. 'Why girls don't do mathematics?'. Today it is becoming clearer that girls do achieve as well as boys in mathematics but choose to participate less. The question now is 'why won't girls do mathematics?' (Willis 1989). There are many in our society, including but not confined to feminists, who share a basic recognition that the position of women in society is one of disadvantage and see education as one of the ways of challenging that disadvantage. They recognise the experience of individual women but locate that experience in the wider structure of society and the recognition that other women share the experiences of the individual woman (Coates 1994). It is this shared experience of learning mathematics that will be explored in this chapter. The processes which exclude and marginalise women and girls in mathematics will be identified and, placing the marginalisation within a broader canvas, see it as a particular case of deeply-rooted social forces which affect other groups in society.

Achievement and participation

This is a controversial area with research indicating different and sometimes contradictory findings. As women's and girls' aspirations have changed over the past decades, so too have the patterns of achievement and participation in mathematics. However, to understand the influences on women returning to study mathematics requires an awareness of the learning context of their initial introduction to formal mathematics at school which may have occurred three or four decades ago.

A survey of over 4,000 grade 12 students in 13 countries (Hanna *et al* 1990) attempted to understand international differences in mathematics achievement, particularly in relation to gender. It concluded that differences in achievement could not be pan-culturally attributed to contextual factors such as the proportion of female mathematics teachers, school organisation, sex-stereotypical thinking, years of anticipated post-secondary education or home support. It suggested that these factors influence achievement in interaction with other societal factors which vary across cultures. What was shown clearly was that gender difference in mathematics varies from country to country (and in particular between two of the constituent countries of Britain). Hence any problems girls, and hence women, have with mathematics is less likely to be a biological than a social construct.

In Britain, at ages where participation is compulsory and hence equal, most girls achieve at least as well in mathematics as most boys. However, there is a disparity between boys' and girls' participation rates in mathematics and girls come out of mathematics education disadvantaged relative to boys. The proportion of females passing mathematics examinations at 16 and 18 years of age in Britain is lower than that of males and from 16 onwards (the age at which mathematics is no longer compulsory) at each decision point the proportion of females opting for mathematics study diminishes relative to males (Ernest 1991). Hence this chapter explores women's achievement but also considers their participation and subsequent attitudes to mathematics.

Gendered patterns of behaviour and achievement can result from attitudes to mathematics. In 1992, 1,471 adult returners on Access courses and 253 individuals inquiring about these courses took part in a survey to establish the attitudes of returners and potential returners to mathematics (Benn and Burton 1994a). These were identified in two ways. There was a direct question, 'Is mathematics a barrier for you?' In addition a Likert Attitude Scale, consisting of ten questions, was developed to test indirectly the deterrent effect of mathematics. Women on Access, whether asked directly or indirectly, showed more anxiety about mathematics than men; and women considering a return to study showed the highest level of anxiety. These results illustrate the experience of many adult education mathematics tutors that years of formal schooling have left many women with less confidence than men in their ability to understand mathematics.

Does gender differentiation matter?

Chapter 8 showed the crucial importance of mathematical competence in all areas of life. It plays a crucial role not only in educational and work opportunities but also in citizenship, leisure activities and the culture of our society. If the reasons for girls' and women's more limited participation in mathematics are educational or social, then it is clearly not just that women are innately disadvantaged. Educational institutions should endeavour to ensure that no students are discriminated against and that wherever possible discriminatory social factors are countered. Even if the problem *were* inherent in females' ability, natural justice would call for extensive remedial action by the institution in the same way that boys' perceived disadvantage in literacy is countered by remedial provision.

Even those who are not swayed by the social justice argument acknowledge that Britain cannot afford to waste the talent of so many of its workforce. Mathematics blocks access not only to those jobs which actually utilise mathematics but also to many more where a mathematics qualification is an entry condition. Mathematics is frequently used as a selector not because it is useful to the course or career but perhaps precisely because it isn't. It is used as an IQ test to separate the intelligent from the less intelligent. So it is a source of inequality. It closes many educational and career opportunities to women and deprives society of the benefit of their talent.

Even if girls do achieve as well as boys in school mathematics, if this process results in long-term alienation then these girls when women are excluded from full participation. To continue this in a supposed democracy is immoral and can only be understood on the grounds of culpable ignorance or deliberate exclusion to serve the interests of dominant groups. It may be argued that some of the findings presented in this chapter are now obsolete and that discrimination no longer occurs. In the market-driven 90s, with its emphasis on individualism and the consequent lack of interest by society at large in collective issues such as equal opportunities, I am sceptical but even if this were true, the women who attend adult education provision were at school in the 80s, 70s, 60s and earlier, and will have had their attitudes and experiences of mathematics shaped by the gender-biased system now discussed.

Who does it matter to and why?

The five ideologies of mathematics education outlined in Chapter 9 have very different perceptions of the problem of gender and mathematics. These will be outlined as a useful framework for our discussion. Ernest (1991: 276–278) suggests that the industrial trainers, the dominant ideology at present, see inequality of women as stemming from the intrinsic hierarchical nature of humanity. They, like the old humanists, see mathematical ability as fixed, inherited and distributed in an unequal way. This view helps to sustain and recreate gender inequalities in society. Technical pragmatists are concerned with the utilitarian problem of the wasted labour resource and hence, although seeing mathematics itself as neutral and value-free, would alter the pedagogy to ensure the attraction and retention of women to this technologically crucial subject. The progressive educators see the problem as located in the individual and seek to correct women's under-achievement and lack of confidence by good classroom practice. The public educators locate the problem in structural and institutional sexism in educational institutions and society. Their approach is anti-sexist rather than pluralist, exposing the gender inequalities in society, educational institutions, teachers and the classroom. They seek emancipatory and empowering education for all.

For real change, the more fundamental approach of the radical public educators is required. However, the ongoing dilemma of the radical is whether to risk marginalisation by advocating a major change in society or take the smaller steps of changing practice within the more accessible locations of the classroom or the institution..

The approaches of the progressive educators and the public educators on mathematics and gender find an echo in those of liberal feminists and radical feminists respectively. Liberal feminists building on the tradition of liberal adult education, described in Chapter 1, link the idea of education as a 'good thing' in its own right to the feminist notion of equality. Hence education is one way of redressing the disadvantage of women by encouraging women as individuals to fulfil their potential and hence equalise the balance between men and women. Although basically individualistic, this approach can challenge structural discrimination at societal level. Radical feminists value the personal, actively challenge the political, and see education as the construction of knowledge rather than the acceptance of received knowledge (Coates 1994).

Radical feminist practice then, has given us the ways to initiate, explore and support change, through the analysis of experience, the fostering of collectivity and the development of new strategies (Spendiff 1987). As with public educators, radical feminists encourage women to see their experiences of mathematics as not located in personal failure but in the positioning of women in society. Discussions around the socio-political framework within which mathematics is learnt are assumed as central to the experience of learning mathematics. This perspective sees education as about change but change as both empowerment for the individual and a move towards a more just society. It is not a comfortable or easy perspective.

An exploration of some of the reasons for gender-related factors in mathematics education

A major part of women's experience of mathematics takes place in general society and the relationship of most women to mathematics is constructed by the receipt of messages about mathematics. Thus it is the content of these messages that constructs the experience of mathematics. Women experience mathematics as an area of competing discourses. They are told on one hand that it is important to learn mathematics whilst on the other that it is less important for women to learn mathematics. They also receive a multitude of other mixed messages which we will discuss below.

Cognitive, genetic or innate differences

Women's more limited spatial visualisation ability is often quoted as proof that men are innately better at mathematics. There are, however, several counter-arguments. First, there is no proof that this ability or the lack of it does affect the learning of mathematics. Even if it does, this difference of spatial visualisation ability may not be innate but due to differential learning experiences such as early play with construction toys and may hence be cause-and-effect with problems in mathematics or may indeed be two different outcomes of the same cause (Fennema 1979). Differences in the experiences, hobbies and extra-curricula activities of boys and girls may influence knowledge of the many contexts in which mathematical ideas and problems are presented and hence their achievement or affinity for the subject.

Whatever, the suggestion that there are innate or genetic gendered differences in mathematical ability are not supported by the results of international studies (Hanna *et al* 1990). Lovejoy and Barboza (1984) suggest that the interpretation of mathematics and science as being a male domain is specifically a Western phenomenon. Western women's problems with mathematics are hence unlikely to be due to some existentially conflicting quality of mathematics and femaleness.

Sexism in society

Feminists argue that our society is characterised by overt sexist beliefs and behaviours, cultural domination and structural institutional sexism. Ernest (1991) suggests that sexism in education and sexism in society interrelate to form a cycle of disadvantage. Girls' lack of learning opportunities in mathematics, for whatever reason, leads to a negative view of their own mathematical ability and reinforces the perception that mathematics is a male subject. A consequence is girls' lower attainment and participation in mathematics. Because of its critical filter role, this leads to restricted educational opportunities and lower-paid, lower-status employment for women. This positioning of women reproduces gender inequality in society and hence reinforces gender stereotyping. This contributes to institutional sexism in education which produces the lack of mathematical opportunities for girls in mathematics. Though simplistic, this cycle does illustrate clearly that the problem lies as much in the socio-political as in the education realm.

Educational

The Girls and Mathematics Unit at the University of London Institute of Education has concluded that underachievement is not the cause of women's under-participation in mathematics, but is due in large part to institutional sexism mediated by teachers (1988). In schools, the story is complex, but could be interpreted to suggest that girls do succeed at what schools ask them to do but right from the start of their education, primary schools and their teachers fail to prepare them adequately for more advanced mathematics. Shuard (1986) comments that 'it would be mischievous to suggest that pupils who pay attention to the teacher's traditional emphasis in primary mathematics give themselves a positive disadvantage for future success in mathematics, but the evidence seems to point in this direction' (quoted in Isaacson 1991: 98). Those that have been well socialised at an early age (girls) may be disadvantaged at a later stage. Scott-Hodgetts (1986), building on the notion of two distinct learning strategies – serialist and holist – suggests that girls may tend towards serialist strategies which are emphasised by primary school teachers and thus reinforced. Boys on the other hand tend to holist strategies and since they, in addition, receive serialist strategy training from their teachers, this ensures they become versatile learners.

Teacher education may partly be to blame. Primary teachers are not forced to examine gender issues in their training and hence are vulnerable to prevailing sexist ideologies when under the pressure of everyday life in the classroom (Skelton 1985). Skelton and Hanson (1989) also show that the assumption on a PGCE course was that boys were better at mathematics despite the existence of all the evidence that, in the early years, girls perform better. Secondary initial teacher education was more likely to address gender issues but this was found to be effective only if dealt with in the practice-based sessions and not the theoretical part of the course.

Teacher attitudes to gender certainly seem to contribute to eventual gender differentiation. Willis (1989), outlining the Australian experience, lists evidence to support the assertion that in the 16 + national tests girls achieve at least as well as boys and possibly better but shows that, when presented with these results, most educators were surprised. This reaction indicates that superior performance from boys is expected. A further factor in gender differentiation may be found in the result of research which shows that in mixed sex classrooms teachers give more attention to boys. Leder (1990) suggests that the results of her and other research give the impression that boys and girls are treated differently in mathematics classes. But of particular interest was that this different behaviour seems to proceed gender differences in performance in mathematics which only become significant at the 16+ level. Hence, though no direct causal link has been established between teacher treatment and performance, teacher behaviours are symptomatic of wider social expectations and beliefs. They reflect and reinforce gender expectations. The teachers in the study were experienced and enthusiastic and hence had not appeared likely to gender-differentiate. The evidence is complex but nevertheless it would seem that treatment of boys and girls in mathematics classes leaves the girls at risk. These gendered expectations may well influence individuals' attitudes to mathematics well after they have left school.

Sexism in the curriculum or in classroom materials

That classroom materials are sexist was shown by an analysis of text books used by the schools participating in The Girls in Mathematics Unit research (Walkerdine 1989: 190–191). They summarised their results as follows:

- In the primary texts, the home and immediate outside world become the salient reference points. Girls/women are equally represented for identification and teaching. They hold positions of power in the primary framework but on the other hand, they are shown repeating processes already learnt. This regulates them as passive recipients of knowledge.
- Girls/women are overwhelmingly depicted in co-operative, helpful and kind roles which mirror the importance of domestic tasks. Also women are generally surrogate mothers. They may be teachers, air hostesses, shop assistants, but their main role is caring for others.

- At the secondary stage, the world of girls and women disappears from the text in everything but the category of identification whereas male presence actually increases to 93.2 percent of the total average.
- Furthermore, any reference to the role of women and girls become devalued by unscientific references to non-mathematical problems.
- Mathematics becomes a masculine province. The expounding of male mathematicians' theories sets readers firmly within a masculine framework.
- The sparse illustrations and gender references contrast markedly with the jolly mixed groups and active participants in the primary texts.

The Girls and Mathematics Unit asserts that both image and text collude to display increasing mastery in boys and a fictional marginalisation and exclusion of girls. The texts do provide sites for identification, but the subject positions they produce consistently represent the feminine as marginal, passive, domestic and failing even at the most trivial of activities whereas the masculine is presented as serious and clever with mastery over the physical and scientific world assured. These are the texts that women who wish to learn mathematics will have used in their time at school. No wonder that for many, school mathematics was an alienating experience.

Affective

Affective factors influencing gender differentiation in mathematics learning have been thoroughly explored by Walkerdine (Walkerdine 1983; Adams and Walkerdine 1986; Walkerdine 1988; Walkerdine 1989). Her powerful analysis is located in the deconstruction of the terms male and female and conducted through 'a history of the present', Foucault's term for the examination of the conditions which produce our taken-for-granted practices. She argues that gender differentiation in mathematics learning is located in the concept of 'real understanding' as opposed to 'rule following' and 'memorisation'. Mathematics of today, like all learning, has two components – the 'knowing that' and 'knowing how' or instrumental understanding and relational understanding.

Walkerdine recognises modern practice in education as propositional learning which produces not just an ability to do mathematics but also ensures the prime requisite of today – the development of a reasoned and logical mind. Attainment in itself is not sufficient: it must be based on real understanding. But definitions of real understanding are located in the prevalent concept of learner-centred learning in which the nature of the learner is defined as active, inquiring and discovering, all the characteristics associated with a white, middle-class male. Boys' play is more active, exploratory and constructive, while girls' is more structured, rule-orientated and inward-looking (Willis 1989; Shuard 1982). What is defined as a properly developing child is recognised by the production of certain required behaviours. The sign of a rationally-ordered child is the growing ability to challenge both the teacher and the discourse of mathematics. Boys' behaviour – difficult, naughty and dynamic – exhibits this required challenging behaviour and hence, even if boys do not attain, they are seen as having potential. Girls, on the other hand, are judged to work hard, neatly and quietly. Though overtly appearing to be exactly what is required of them by the teacher's words, this behaviour and its consequent learning is not valued by the teacher. Consequently, when girls do attain, this is not judged as proper or real learning. The result is that girls who succeed in a way that does not fit our image of an active, inquiring child are considered to be not really capable. Their success must be explained away. Girls do very well in primary schools but this is not seen as real learning, they have just worked hard (Walkerdine 1983).

High-achieving girls are never accorded the accolade of 'brilliant'. On the other hand, boys' early failure is no failure. They may be termed bored or slow developers but they have potential. This undervaluing of girls' success in mathematics in school is arguably communicated to both the girls and boys either implicitly or explicitly. This reduces girls' confidence in future success. It also conveys conflicting messages. Girls are successful on a day-by-day basis but can never change their teachers' perception that this success, because it is due to rule-following not rule-challenging, does not indicate real mathematical aptitude. Their voices are less likely to be heard in the classroom. As Fennema *et al* (1990) showed, even gender-aware teachers have not eliminated inequalities in their classrooms. These teachers are aware that girls are not learning mathematics as well as the boys but still hold differential beliefs about girls, boys and mathematics. Many of the behaviours perceived as different for girls and boys are the autonomous learning behaviours so valued by teachers as important in the learning and use of mathematics.

The dissemination of the research outlined in this section may have affected classroom practice and differences in achievement and participation in mathematics are narrowing. But women learning mathematics now will have experienced their mathematics learning in the gendered classrooms outlined above. All these points may help to explain girls' gradually removal from the mathematics scene and women's subsequent alienation from mathematics. Even more importantly, adult educators could usefully consider whether this concept of 'real understanding' with concomitant undervaluing of female behaviour exists in the adult classroom.

Gilligan's research (1982) in the field of women's moral development provides a link between mathematics avoidance and mathematics pedagogy. Gilligan identified two forms of reasoning which are gender-related if not gender-specific. These are separated reasoning, often identified with men and hence normalised, which is characterised by objectivity, reason and logic and connected reasoning, the 'other' voice often identified with women and hence devalued, which is characterised by subjectivity, intuition and the importance of relationships. In the connected approach, mathematics as a subject is seen as intuitive, connected and constructed. In the separated approach, it is seen as finished, absolute and predigested. This does not involve the learner in any deep sense nor does it help the learner to construct meaning in their own lives and of their own experiences. But to move from separated to connected learning requires initiative, independence, challenge-and risk-taking just the attributes of the autonomous learner which has already been argued is developed in boys but not in girls. This concept helps us to understand yet another force that, whilst not necessarily affecting girls' ability in mathematics, could contribute to their alienation and hence their voting with their feet. The result for women learners, who so often return to study to obtain a clearer understanding of their lives and experiences is that, far from thinking mathematics could help in this process, they see formal mathematics as a discourse in a different and alien tongue, a discourse of certainty and objectivity set outside any real social context. Not surprisingly, they acknowledge its power, mourn their lack but see it as 'other'.

This is illustrated by the comments of women returners to study that involved mathematics (Benn and Burton 1994b). Over half felt that they had experienced discrimination and discouragement in their earlier study of mathematics. They reported negative experiences with mathematics at the secondary level which had left many of them with high levels of anxiety and low levels of confidence.

> I believed I could not do mathematics. I was frightened of mathematics. The mathematics teacher was not interested in students like me.

Perceptions of the usefulness of mathematics are also gender-differentiated. Isaacson (1989: 188) has suggested two theoretical constructs, that of coercive inducements and double conformity, to aid insight into this aspect of the gender and mathematics issue. The notion of coercive inducements starts with the premise that women are not prevented from taking non-traditional roles in society or subjects at school. They themselves choose but under inducements so powerful that they are effectively coercive. Although this is a free choice, the perceived rewards of entering a more socially acceptable subject or career option are too strong to resist.

To this is added the notion of double conformity (the dilemma of any person who has to conform to two inconsistent sets of standards or expectations). Research carried out (Broverman *et al* 1970) into people's views of a mentally healthy, mature, socially competent adult, man, or woman showed that the characteristics of a normal adult and a normal man correlate very closely while those of a normal woman are very different. This implies a conflict for women who wish to be both a normal woman and a normal adult. Girls or women who wish to study subjects such as mathematics or science which are identified as in the male domain or have careers in 'male' occupations will have to conform at the same time to two mutually-exclusive sets of criteria. This aids understanding of the obstacles confronting women and girls who wish to study male defined areas of study. It is not surprising if coercive inducements pull and double conformity pushes women out of mathematics.

What can be done to overcome gender differentiation in learning mathematics?

Women returners to mathematics have their earlier school experiences located in a system which pathologised women and normalised men. The system evolved from a complex mix of historical, social and political factors. Mathematics has played its role in the systematic disadvantaging of girls and women. Whatever is to be done about it will not be simple or quick nor can this structural issue be confined to the learning of mathematics. Nevertheless, educators can contribute by means such as meeting regularly to discuss strategies, pedagogy and curriculum. In the area of adult education, many tutors work in relative isolation, some with no formal training in adult education themselves. Groups such as the Adults Learning Mathematics Research Forum provide a space for researchers and tutors interested in gender issues to share knowledge and experiences and develop alternative approaches.

Though awareness of gender differentiation has increased, this is an ongoing issue in school mathematics. It is certainly an issue for the women whose schooling in past decades has left them inadequately prepared for a changing future. The comments of three women returners given below show their impressive motivation to obtain the GCSE mathematics qualification essential for entry to teacher education. They provide the impetus for adult mathematics educators to continue with the very real political struggle to change the experiences of women like these and endeavour to ensure that their daughters and grand daughters need not endure the same:

> I would have just carried on and kept trying – I mean I took it twice – I took it at school and I took it when I left school and failed miserably.

> I am determined, I am motivated but if I haven't got that ability then I'm lost and I think that's what really frightens me and I think the mathematics requirement is a big barrier for me.

> When I was told I needed mathematics it was quite a challenge to actually go back and say right I'm going to do this. And I did sit it and actually got a D – it wasn't good enough. I needed a C . . . I am highly motivated because there's a goal – I want to go into teaching.

The idea of gender planning, where the aims and objectives of the course, the programme and the institution are considered in relation to women as well as men, could be incorporated into all levels and forms of mathematics provision (McCaffery 1993). These aims and objectives would then be monitored and evaluated in relation to their impact on both sexes. The purpose of gender planning is to incorporate the differing and distinct needs of women into the mainstream of the planning process and thus of the delivery process. The process requires a gender analysis of:

- the economic and social context in the area in which the provision will be offered,
- the social construction of gender at the household level within the cultural, religious and communal traditions of the target groups,
- both practical needs (*eg*, child care) and strategic needs (ways of challenging traditional gender roles),
- a gender diagnosis of the institutional/educational framework,
- if necessary, the development of educational programmes with gender-specific aims and objectives.

The practices we are suggesting fit well into the feminist framework outlined above. Some are evolutionary, according well with the liberal feminist approach. Others are more emancipatory and in line with the radical feminists.

One approach advocated by both these groups is that of women-only groups. In mixed groupings, men tend to dominate, women are less likely to challenge men or male assumptions and are not given space to explore their own perspectives. Mixed groups are more competitive and less supportive. Belenky *et al* in their influential book *Women's Ways of Knowing* (1986) argue that a learning environment is needed which gives women a space for themselves where their voices will be heard and women's ways of knowing, thinking and preferred styles of learning will be acknowledged and valued. The ideas of Mezirow outlined in Chapter 1 illustrate the process of change that can occur for women in the (hopefully) secure unjudgemental environment of women-only provision. These ideas need to be extended, however, to recognise that women's re-entry into society after their learning experience may be in anger at the realisation of the injustice that they have experienced and hence may result not in an easy reintegration but perhaps a struggle to combat the oppressive hegemony.

Not all feminists advocate women-only provision. As Malcolm suggests (1992), there are inherent dangers in this practice. Institutions, if constantly reminded that women are different from ordinary students, may continue to organise themselves on the assumption that ordinary students are not women. The strategy of claiming that women require different courses has serious implications for the majority of women students who attend mainstream provision, since it poses no challenge to existing practices. If these practices were to be significantly changed, then arguably there would be no or less need for women-only courses.

Women need to be told that they are capable of doing mathematics, even if to the teacher this may seem obvious, and the potential uses of mathematics in women's lives need to be drawn out. Tutors need to be constantly aware of the legacy of school years. The need to provide a relaxed environment is essential as a site for discussion and for collaborative, co-operative but challenging and stretching learning. The importance of group work is absolutely crucial. The feature which most distinguished adult educational experiences in mathematics from that of school has been identified as the strong sense of peer group identity and mutual support. Students receive considerable emotional and practical help from the rest of the group (Benn and Burton 1994b).

The elimination of overt and covert sexism from text books, other curriculum materials and examinations and the widening of the range of applications from sport, ballistics and war to everyday or socially valuable examples (appliances for the disabled) would help. The majority of textbook authors, curriculum developers and tutors are male and since the choice of examples and contexts for mathematics problems will naturally arise out of the teacher's experience and interests, this can lead to systematic distortion and restriction of context choice. Bias in the choice of context can lead to systematically putting certain groups at risk. Previous experience and knowledge can lead to an empathy or otherwise with a context and hence advantage or disadvantage in the solution of the problem (Verhage 1990). A wide choice of contexts would help whilst, at the same time, widening the cultural knowledge of all learners. Student groups generating their own problems and contexts may ensure diversity and a more representative sample.

Many of these approaches, including collaboration and sharing of control, co-operation rather than competition, autonomous and self-directed and self-controlled learning behaviour, and a pedagogy which allows for and learns from error and conflict, are already being developed and used by gender-aware teachers. The returner survey mentioned earlier showed that much of the considerable and commendable success in overcoming women's fears of mathematics can be attributed not just to the level of motivation of students and peer group support, but also to the approach taken by tutors (Benn and Burton 1994b). The main elements of their teaching are encouragement, understanding, patience, the removal of the often difficult and disabling pressures of time and the recognition that mathematics needs to be taught in context and have a relevance to real life and other parts of the course. Tutors noted that the involvement of students and tutors in free discussion and dialogue in a supportive atmosphere helps students develop confidence. Most noted an urgent need to build confidence by showing that it is acceptable to be wrong and by placing the emphasis on methods rather than answers; to develop a positive attitude to mathematics by encouraging students to take ownership of mathematics by exploring and enjoying numbers. The following comments illustrate the effectiveness of these approaches:

> An excellent mathematics tutor . . . has made mathematics enjoyable for me.

> I was on the Access BEd course I found that the different style of teaching refreshing, and I now enjoy mathematics. I am now doing well.

> Just a drastic change in teaching techniques . . . demonstrated my enthusiasm to learn mathematics.

Conclusion

Many girls and women do succeed in mathematics. Many girls and women do better than most boys or men and many men and boys do worse than most girls and women. Many women do use mathematics in everyday life and many returners to education do study mathematics. But many women are 'subtracted' from mathematics and the alienation remains with them throughout their lives. This is not a problem of or with girls and women. It is the obligation of a democratic society to ensure that women have real access to the empowerment that mathematics endows. This calls for rigorous and demanding mathematics education, embedded in real world concerns and people-dominated contexts illustrating the fallible, non-absolute nature of mathematics, its cultural and social power as well as its utilitarian function for work. It requires a greater understanding of concepts such as 'real learning', student-centred curriculum, stereotyping notions of success and failure, and women's empathy with connected learning rather than the separated version of mathematics. It calls for the explicit raising of social, political and cultural issues in classroom to enable both men and women to see the way that cultural conditioning affects perceptions and how these perceptions affect achievement and attitudes. Linking rationality and reason with the affective and the social would contribute to the elimination of gender differentiation in mathematics learning.

The answer to the question 'why won't women do mathematics?' is not genetic but social. It lies in society's tendency to normalise male behaviour and pathologise female behaviour. There are contributory factors from the institutions in which mathematics is learned and from teachers but these individuals have been positioned just as much as the students. Inequalities are structural and serve the interests of powerful groups in society. This means that any effective moves to change the *status quo* will probably be attacked and marginalised. Teachers in compulsory education must bear the brunt of any attacks but adult education still has, even if to a reducing degree, greater anonymity, autonomy and freedom. It can continue to do what it has always done, form pockets of resistance and fight injustice from the sidelines.

Perhaps even more than school teachers, adult educators' constant encounters with women who have been alienated and often distressed by their formal mathematics experiences will give them courage and energy to continue the struggle. But the struggle may be personal as well as political. To address equality issues is to challenge at a personal as well as an institutional or societal level. Confronting inequality involves the tutor in critical self-evaluation. Personal assumptions, attitudes, expectations, behaviour are all called into question and may require change in not just the learner but also the tutor (Skelton and Hanson 1989).

Kaisser (1996) identifies five phases of mathematics in relation to gender. First is *womenless mathematics* which was common until the 1970s. Women's role in mathematics was silence and exclusion. Next came *women in mathematics*, with women entering mathematics if they conformed to male attributes.

Then, in the 1980s, came *women as a problem in mathematics*, with the emphasis on intervention projects. Here, the problem lies with the learner, not the subject. The fourth phase, which is arguably the one we are in transition towards, sees *women as central to mathematics*, shifting the blame for women's lack of involvement in mathematics from women to the system. The fifth phase, as yet over the horizon and hence not clearly defined, might be *mathematics for all*, a reconstruction of mathematics as a connected and constructivist discipline.

To date, gender activities in mathematics education are on the margins. Despite the efforts of many educators, the focus has been on access to mathematics rather than any substantive change in the curriculum. Currently equity can be seen as pretending not to notice, as removing consideration of gender, class or race on the grounds of equal opportunities. However, if we remove these definitions from our discourse, then we cannot challenge prejudice when it occurs. The curriculum is not innocent. Socio-economic, gender and other factors still affect participation and achievement. To change this situation, this needs to be raised and discussed openly in society in general and in the mathematics classroom in particular (Willis 1996).

But perhaps translating 'why won't women do mathematics?' to 'why don't women do mathematics?' is misleading. Many women do mathematics and they do it frequently, capably and well. But this may be workplace or everyday mathematics which is seen as a limited and impoverished version of the real thing: a mathematics without a name and not to be valued. This will be discussed further in Chapter 15.

Chapter Thirteen

A class apart

There can be no doubt that Britain has long been and remains characterised by deep and structural inequalities. In particular, it is a peculiarly class-ridden society. Gross inequalities of wealth persist and indeed in the 1980s and 1990s such disparities widened considerably (Benn and Fieldhouse 1990; Taylor 1986). However, the concept of class has fallen into disuse in the discourse of education due to several factors. Firstly, the term 'working class' is itself suspect and can be seen as pejorative, divisive, patronising and irrelevant in terms of adult education provision (Thompson 1981). Secondly, the educational agenda is now defined around the needs of the workplace and 'Britain plc' rather than the cultural context of the working class and its incompatibility with formal education. Thirdly, it fell victim to the politics of equality of opportunity which dominates the educational agenda.

'Equal opportunities' was seen by many as a radical attempt to change the educational agenda. However, it could alternatively be seen as the promotion of individualism and individual rights as the cornerstone of society; the equal opportunity to compete in this as opposed to a more equal society. There was little recognition of the difference between equality of opportunity and equality of outcome or that some groups in society were more able to take advantage of the same opportunities. The present categorisation is that of disadvantage which is now so pluralistic (gender, race, age, rural, educationally disadvantaged and disability) that it means almost nothing. The new problematic refers generally to a wider client group with more diverse needs and ambitions (Watson 1993). Clarke (1993) argues that, in many academic circles, class became synonymous with white male values and hence politically incorrect. The exploration of 'other' now locates on race and gender and rarely includes reference to class.

Here we argue that the concept of social class is relevant to our discussion of educational opportunities for adults to learn mathematics. Jackson (1981:104) defined working class as:

> an aggregate of low status individuals . . . whose low status is defined by social and economic criteria: employment (or lack of it), income, educational background, job opportunities, general access to resources indicated by area of residence, ability to choose residence, lifestyle etc.

There remains a working class culture with distinctive roots with three main organisational sites developed in the 19th and 20th centuries: consumer co-operation, trade unionism and the Labour Party. The wider culture includes the pub, the football club, the friendly society and a host of institutions which together with the distinctive patterns of family life add up to a class culture. By definition, the working class enjoys less than full citizenship in society. Education is one element in the movement aimed at changing this state of affairs by ensuring full participation of the working class in a political, industrial and social democracy. But this has not succeeded and differentials in attainment continue to exist between people of different class origins despite all endeavours to remove formal inequalities and the considerable expansion of opportunities at all levels of education (Halsey 1975).

Participation and achievement

The evidence seems incontestable. Adult education generally attracts a small and socially discrete section of the population, the middle and lower-middle class who have already experienced a fair amount of educational success and who are not principally motivated by vocational requirements. This is particularly true of the WEA and universities provision but even the local authorities, who provide most of the mathematics education for adults, have a very similar student body which differs only by being slightly older and more likely to be lower middle class than middle class. The NIAE study *Adequacy of Provision* (1970), the ACACE survey (1982a), the 1987 FEU survey *Marketing Adult Continuing Education.* the Munn and McDonald survey in Scotland (1988) and McGivney (1990), together with smaller local surveys, all show that the adult education service has chiefly benefited those in the higher social classes. Adult education of all kinds and at all levels recruits disproportionately from non-manual rather than manual occupations. The ACACE report (1982b) observed 'the phenomena *(sic)* of social class as it affects education in this country is extremely powerful'. A comprehensive review of the literature showed that in Britain as elsewhere in the USA and Europe working class people, particularly women, are massively under-represented throughout post-school education (McGivney 1990). The main characteristic of non-participation appears to be social and economic deprivation.

Does class differentiation matter?

A strong association exists between social background and educational performance (DES 1988, Uden 1996). Adults from lower socio-economic groupings have often achieved less well in school than those from the middle and professional classes. However, working class people often perform competently in jobs that require technical knowledge and skill. If these cognitive tasks can be carried out in the workplace, whether using formal mathematics or an ethnomathematics, then this has educational implications. Different contexts such as work, home and school are characterised by different practices and related sets of terms and meanings which we have identified in Chapter 10 as discursive practices. Individuals are put into positions by the practices in which they are engaged as well as structurally by their social class origins. The discourses of workplace mathematics are not the same as those of family mathematics, school mathematics or formal mathematics. They may be particularly disparate for the working class whose home and family does not fully resonate with that of school and the wider society (Evans and Harris 1991).

This is supported by evidence from the investigations carried out by Carraher and Schliemann (1988) into the mathematical knowledge of children from both lower and middle class socio-economic groups in the north-east of Brazil. It was anticipated that many of the students from the lower socio-economic groups would fail in mathematics due to poverty and consequent malnutrition. At the end of the year of the study, as expected 32 per cent of the lower class pupils who had participated in the study failed mathematics as compared to 2 per cent of the middle class youths. However, cognitive assessments were also carried out which showed that there were no substantial differences between the middle and lower class children in mathematical skills or understanding after the year of mathematics schooling. This raised questions about the neutrality of schools in assessing the cognitive competence of children.

In America, poor performance has been traced directly to both blatant and subtle discrimination and extreme poverty (National Science Foundation 1983): when students from the lower socio-economic groups are exposed to a good learning environment, they perform as well as any. So low achievement norms do not reflect ability; they reflect a lack of preparation and early exposure. Further research in America has shown that basic mathematical thought develops in a robust manner among lower and middle-class children. Children enter school with some mathematics knowledge, a desire to learn and the ability to perform adequately (Ginsburg and Russell 1981). But from the very early school years children are socialised into the roles that their socio-economic background determines (Apple 1982). Teacher expectation for working class or poor children is low, resulting in poor performance which is exacerbated through the years of schooling.

Streaming can severely disadvantage working class children and further contribute to low expectations by staff and pupils and continued deterioration of performance. If further streaming is on the basis of achievement rather than potential, this forms a trap which cannot be escaped. A differentiated curriculum completes the process through low-level practical mathematics for the working class for eventual low-level occupations and high-level theoretical mathematics for the middle class for eventual high-level occupations. This will result in active or passive alienation and resistance for the former and autonomous critical citizenship for the later. This background of mathematics education for working class will be brought into the adult education mathematics class.

An exploration of some of the class-related factors in mathematics education for adults

Practical factors

Various explanations have been made for the failure of the adult education movement to make an impact with the working class. Cost is a major perceived barrier and several national and regional surveys showed that sharp fee increases which have characterised some recent adult education provision have substantially affected participation by those from the lower socio-economic groups (McGivney 1990). Barriers to participation for the working class include time constraints especially for those doing part-time or shift work. Many adult education courses assume students possess a secure financial basis and plenty of time to study. Many working class women in particular have to work, function as adults and have childcare commitments without the middle class network of support for study (Clark 1993).

In addition, education is not seen as part of reference group norms. McGivney (1990) suggests that many people in working class occupations are hostile to the education system and hence to education in general which is particularly important as reference group attitudes and norms exert a powerful influence over attitudes and behaviour within this group. Effective recruitment therefore needs to be located in working class culture and perhaps through the workplace or word of mouth.

The learning programmes that have recruited manual workers have generally been informal student-centred and democratically-run. Surveys consistently reveal that the expressed learning interests of male manual workers are orientated exclusively towards practical and vocational skills. Hence, linked skills programmes involving both basic and practical skills can be particularly effective in attracting male manual workers. It would certainly seem as if the learning of mathematics could be integrated into such a programme. The problems used could be linked to the desired skills whether the practical skills of carpentry or the critical skills of statistics in, say, labour studies.

The value system of adult education, located as it is in institutional practice, seeks to produce quality through conformity in structure, method and organisation. This is alien to the working class lifestyle, with its emphasis on small, informal group relationships and social contacts. The formal role of the teacher is also alien to this lifestyle (Lovett 1975). The sequential and discipline-based nature of much formal adult education is often inappropriate for those who seek learning that is immediately relevant. There is a further difficulty that student-centred, needs-meeting learning requires the ability in the learner to recognise, formulate and articulate the mathematics required and in the tutor to interpret and translate these needs. Many working class people find adult education centres unwelcoming places and are deterred by the timing and siting of classes, lack of creche facilities and chaotic and insensitive procedures at enrolment time. The lack of resources of the working class is an additional enormous disadvantage (Tuckett 1991).

Social factors

A further barrier is the middle class nature of adult education itself. In times of recession and the current preoccupation with the market, the tendency is for adult education to service those who most readily come forward. As we have shown this tends to be those had reasonably good and positive experiences of initial schooling: the middle classes. Schuller effectively describes this as 'the phenomenon of second creaming: an increase in services principally benefits those who just failed to profit from what already existed, leaving others relatively worse off' (1978: 25). People who have 'failed' the school system do not wish to repeat the failure. If school has reinforced cultural constraints arising from cultural and social class divisions, then many working class people may be programmed to feel that education is not for them. Voluntary learning is perceived to be part of the cultural pattern of higher socio-economic groups. The common language, shared experiences, implicit assumptions and agreed frames of reference of the middle classes establish the boundaries.

Simply put: for the middle class adult education appears relevant; for many working class people it does not. Research in an inner city school supports this class, rather than ability, divide (Walker 1988). It indicated that the more academically successful students were those whose culture converged with that of the teachers and that the strength of this 'intercultural articulation' determined the likely success of educational outcomes. The main cultural divergence was identified as that between the middle class teachers and the working class students. Where divergence was great, then success could only come from teachers developing cultural 'touchstones' that expanded the range of common interest between the two groups.

The view that mathematics is hierarchical in nature

There is a widespread assumption that there is a fixed linear hierarchy of mathematical ability from the least able to the most able. Every person can be assigned a position on this hierarchy and few shift their position during their lifetime. An alternative assumption is that teacher expectations and stereotyping become self-fulfilling and curriculum differentiation exacerbates existing differences (Ruthven 1987). Labelling individuals as 'mathematically low attainers' ensures low attainment. Streaming by ability, which is widespread in the teaching of mathematics, has the effect of labelling and thus affects achievement in mathematics. Ability stereotyping in mathematics is due not only to measured attainment but also class factors (Meighan 1986).

Vygotsky (1962) argues that a learner's capabilities are not fixed but can be extended through social interaction. He acknowledges individual differences in mathematical attainment but suggests that the cognitive level of student response in mathematics is determined not by the 'ability' of the student but by the skill with which the teacher is able to engage the student in mathematical 'activity'. This implies a need for a pedagogy which relates to students' goals and culture. Students labelled as 'mathematically less able' can dramatically raise their levels of performance when they become engaged in socially- and culturally-related activities in mathematics (Ernest 1991).

The ideologies of mathematics education

Using the five ideologies of mathematics education outlined in Chapter 9 will provide a useful framework to this part of our discussion (Ernest 1991). The most influential group in mathematics education are the industrial trainers with their strong belief in the hierarchical nature of both society and mathematics and who want mathematics education, like all education, for social reproduction. This implies a differentiated hierarchical mathematics curriculum with basic skills training for the working class and higher mathematics for the middle classes. Lower ability students are offered a low-status mathematics curriculum to prepare them for low-level occupations. This accords well with the other influential grouping, the old humanists, who are interested in the purity of mathematics and the learning of the subject for its own sake. This group is interested in the provision of high-status theoretical mathematics for high-ability students who will enter high level occupations. These two influential groups advocate the same hierarchical curriculum, though for different reasons.

The technical pragmatists, on the other hand, encourage the upward social mobility of the technologically skilled and are more meritocratic, concerned with mathematics as an important element of education for employment. However, by not locating education in a wider social political framework, this approach tends to loosely recreate the hierarchical structure of society. This is also true for progressives who concentrate on improving the lot of the individual rather than endeavouring to change society. Neither of these two groups question the socialisation of the working class to different educational expectations or the hidden curriculum which validates the middle class culture whilst marginalising that of the working class.

The radical or public educator perspective in complete contrast locates all education in a social context seeing it as a means of achieving social justice. This group actively endeavours to break the cycles of reproduction of the other ideologies through emancipatory education. Here mathematics ability is perceived as not fixed but fluid, leading to a theory of mathematics education which is flexible and sensitive to the social context of the learner.

In Britain today the predominant groups are the industrial trainers with the old humanists and technical pragmatists being influential in compulsory education. The most influential approaches in adult education are those of the industrial trainer because of their hold on the finance and the progressives through their influence on the belief systems of the tutor body. The latter, by its concentration on the individual, provides a caring student-centred pedagogy which allows some learners social mobility through permeable class barriers.

There is also a small group of radical adult educators who, seeing mathematics education as contributing to active and critical citizenship, advocate a questioning and discursive pedagogy to ensure mathematics education is a form of emancipation and a contribution to social justice.

What can be done to overcome class differentiation?

To locate the concept of disadvantage in personalised or individualised explanations is to divert attention away from the more fundamental examination of the structural causes of poverty, inequality, educational divisiveness in our society and the vicious circle of poverty, poor educational performance and limited life chances. The problem is not located in the socially and educationally deprived but within the class divisions in our society. These are reflected in the value system inherent in much adult education.

The view of cultural deprivation and disadvantage which has been very influential in education generally and expounded in the influential Russell report *Adult Education: A Plan for Development* (DES 1973) can be criticised on the grounds that everyone has a culture even if this is different from the mainstream. Hence cultural behaviour is not deficit but difference and different cultures are rated differently in our society. The behaviour patterns of the working class are considered inferior to those of the middle class. Consequently, working class people may be rejecting mainstream mathematics as a part of the rejection of other mainstream culture (Thompson 1981). They manage by the development of their own ethnomathematics. The equating of educational disadvantage with deficit or deprivation leads to the concept of compensatory education. Viewing disadvantage as a result of structural inequalities, on the other hand, results in a perceptive transformation and the requirement for emancipatory empowering education.

This theory is difficult to relate to everyday practice. We have previously argued that it is immoral to not provide the student with the skills and qualification in mathematics for which they came. To locate study solely in the ethnomathematics is to deprive learners of the knowledge that they need if they are to transform their social and economic condition. To concentrate on political theory is to ensure the class votes with its feet. The answer seems to lie in high expectations, locating learning in its social context and genuinely using the student's ethnomathematics as the basis to move into the discourse of formal mathematics. Work with low-income Mexican-American students in America using this approach has been successful by instilling in the students a pride in their Mexican mathematical heritage and high expectation that the students will work hard and do well (Zaslavsky 1990). This approach has also been successful with Access students in Britain (Benn and Burton 1993).

The ideas of Freire can be usefully re-examined here (1972). He saw the teaching of literacy as a radical process of self-discovery and increasing social consciousness resulting in the realisation by the poor that contrary to the myths and expectations imposed upon them, their action can influence the world. At the same time as learning content, the learner must be engaged in a critical analysis of the social framework in which s/he lives. It is important to be clear that the prime purpose of this learning experience is not to raise awareness. It is for the adult to learn mathematics. However, the underlying premise is that without critical awareness, 'other' students will not become engaged in what to them is alien knowledge. To refocus the knowledge, Freire advocates the need to start where the people are, using their knowledge and their culture. Similarly in Britain, if adult education is to succeed with the working class, it must use cultural forms familiar to those involved. This is to fulfil genuinely the adult educator's claim to start where the student is and value the student's experience as an integral part of the learning process. The social experience of the learner are already valid and significant and should be reflected back to them as such (Bernstein 1970).

An example of a situation where the 'correct' solution of a mathematics problem is clearly linked to middle-class norms and expectations is given by Ladson-Billings (1995). The problem given to a group of American youngsters was as follows:

> It costs $1.50 to travel each way on the city bus. A transit system 'fast pass' costs $65 a month. Which is the more economical way to get to work, the daily fare or the fast pass?

The white middle-class youngsters suggested the daily fare was cheaper ($1.50 each way for approximately 20 days a month would be $60). In contrast, many of the inner-city youngsters saw the problem as much less clear-cut. Their experience was that people often had several low-paying part-time or full-time jobs so might need to take the bus more than twice a day. They also suggested that those without a car might use the bus for other reasons apart from commuting such as going to the cinema or visiting friends and family. These factors might all conspire to make the pass better value for inner city people without a car. Here context, rather than mathematics on its own, determines the answer.

There is a need for learning contexts which trigger the imagination. The working class are capable of using concepts but these need to be strongly linked to specific contexts. This does not imply that working class learning should be restricted entirely to working class language and culture but that these should be valued and used as a touchstone to develop a mathematics with wider applicability. They can be used as the stepping stone to move from the ethnomathematics into the discourse of formal mathematics. But if the learner is to use their knowledge and experience as a means of linking practice with theory and concepts, there is a need to re-evaluate what we as educators and we as society count as knowledge. There is a need to re-examine prevalent assumptions that mathematics is absolute not relative, abstract not concrete, context-free not context-bound as discussed in Chapter 3. A move to a more constructivist approach to mathematics allows the knowledge and social experience of the working class learner to be reflected back as valid and significant;

the mathematical knowledge hidden in the family, work or community to be recognised. From this the learner can then move to a wider awareness.

Most tutors are themselves middle class. This cultural difference between tutor and working class student is a further factor in participation and achievement. Crucial to working class culture is the element of equality and common interest implicit in the term 'solidarity'. The prevalent concept of adult education as needs-driven implies students with needs and tutors as needs-meeters. The tutor may perceive these needs through a filter of middle class values. It is the responsibility of the tutor to be aware of these issues and ensure that the relationship between tutor and student is built on demonstration of solidarity and on principles of equity and genuine mutual respect.

The model of 'solidarity' can sometimes usefully supersede the deeply-ingrained model of 'service' (Head 1991). Learning is then a dialogue where the learner's responsibility is to initiate the dialogue and the tutor's to move beyond the boundaries of the already known into the unknown which still has to be discovered, understood, mastered and controlled by the learner. The starting point may vary but the end result is confidence in the discourse of academic mathematics, with the same intellectual demands made on working class learners as on any other. The underlying assumptions are that low motivation and achievement are closely related to alienation; working class adults are capable of undertaking sustained demanding education if it is seen as relevant to their needs (Lovett 1988); and the mutual dialogue which incorporates class interests changes the opportunity to learn. The result of such a pedagogy is that the learners acquire the skills and qualifications in mathematics that they need for social and economic advancement but also the critical awareness required to change or attempt to change the restrictions acting upon them.

This is not an easy process for the adult education tutor. Most education institutions are committed to institutional-based and qualification-orientated provision, linked ideologically and structurally to existing patterns of provision and to the dominant culture and its assumptions. Nevertheless the prize of greater social justice is worth the struggle and as 'others' such as the working class participate more in formal mathematics, our ways of perceiving mathematics will expand. This will be liberating for all of us.

Chapter Fourteen

Adding race to the equation

> Children who need to count and multiply are being taught anti-racist mathematics, whatever that may be.
> (Margaret Thatcher, Conservative Party Conference 1987)

This chapter builds on the premise that the term 'race' is a social categorisation of people based on the tendency of some physical characteristic to be distinguishable and that the term 'black' is a political rather than descriptive term (Leicester 1993). Hence racial differentiation, though obviously overlapping, is different from cultural differentiation and racism is a perspective of deviance not difference. Significant racially-based social, economic and educational disadvantage does occur in Britain, and it is African-Caribbean and Asian people who most experience this racial discrimination and disadvantage.

The Swann Report (1985) identified two different forms of racism: individual racism which is premised on prejudice, and institutional racism. The latter is defined as the way in which a range of long-established practices and procedures which were originally devised to meet the needs of a relatively homogeneous society may (perhaps unintentionally) work against minority groups by depriving them of opportunities open to the majority population. For example, if arrangements for the election of school governors fail to take account of the need for minority ethnic group representation, then this affects democratic rights. Other examples of institutional racism include lack of facilities for people who do not speak English, and culturally-biased assessment procedures and tests. Racism can appear in the education of adults learning mathematics at individual and institutional level and must be countered at both levels. Before exploring these issues further, we will briefly summarise some of the on-going debate concerning racist education, anti-racist education and multicultural education.

The difference between a racist and anti-racist education system is as follows. The racist version consists of mainly white students, an ethnocentric curriculum and teaching approach which does not explore the oppression experienced by black people. The anti-racist version, on the other hand, involves black students in proportional numbers as well as blacks-only provision, an anti-racist, pluralist curriculum and an anti-racist pedagogy. However, it is important to note that such linking of 'anti-racist' and 'pluralist' is contested. Multicultural education is fundamentally liberal with goals of intercultural understanding, limited social reform and a pluralist curriculum. It has a focus on 'other' cultures, implicit anti-racism and social and cultural aspects. On the other hand, anti-racist education is fundamentally radical, with goals of radical social change, the removal of institutional discrimination and an anti-racist curriculum. It has a focus on the racism of the dominant culture, explicit anti-racist teaching and economical and political aspects. For many, these are ideologically irreconcilable. Though recognising this argument, the remainder of the chapter supports writers such as Grinter (1985) and Leicester (1993) in the need for pluralism in the curriculum *and* the elimination of discriminatory institutional practices. The dilemma as to whether to press for modest attainable goals (here of multiculturalism) or more radical ones (here of anti-racism) is one echoed throughout this book. Each adult educator must move to their own considered opinion. What is clearly not acceptable is the continued acceptance of a racist education system.

Participation and achievement

Differences in achievement between different ethnic groups becomes visible during statutory education. The report of the Committee of Enquiry into the education of children from minority ethnic groups in Britain concluded that the achievement of pupils from certain ethnic groups were significantly below average levels at 16 (Swann 1985). This was supported by a study of 3,000 pupils in four urban areas which found that children in some, but not all, black groups scored substantially below average on entering secondary school (Smith and Tomlinson 1989). Even bearing in mind regional and socio-economic factors, it is arguable that many minority ethnic group students are not offered an equal opportunity to achieve and learn in British schools.

When education becomes voluntary as in adult education, this difference in achievement can be exacerbated by a difference in participation. McGivney (1990) argues that largely absent from all types of continuing education are certain specific groups including unemployed adults, some rural populations, immigrants, the aged, urban poverty groups, unemployed and under-employed workers with little education, unskilled and semi-skilled workers, some groups of women and people with linguistic problems. Individuals from minority ethnic populations often fall into these non-participant categories. Their communities usually have higher levels of unemployment than the national or regional norm. They manifest high levels of poverty and social deprivation, are often associated with low levels of formal education, are largely concentrated in unskilled and semi-skilled jobs and sometimes have linguistic problems (Jowitt 1995).

Economic restructuring has brought in its wake social marginalisation with ongoing high levels of unemployment, increasing social deprivation and social divisions. Much of this has been concentrated in minority ethnic communities. Proportionally, they are less likely than the general British population to be in paid employment, although a higher proportion are of employable age. However, Sargant's study (1993) of the participation in education and training by adults from minority ethnic communities showed that, despite these disadvantages they have a high level of participation in study, particularly in informal learning. For those in work, most education and training opportunities were not provided by current employers and there is strong evidence of unmet demand particularly for vocational subjects. Sargant estimates that five per cent of those from minority ethnic groups would like to learn mathematics. The continuing provision of community-based provision is important for minority ethnic groups, and particularly for women. All this suggests that minority ethnic groups are an important sector for those interested in helping adults learn mathematics.

However, the disadvantage incurred by these groups in initial education may result in a lack of confidence which inhibits later interest in participation. In a national survey of adult returners on Access courses (Benn and Burton 1994) 7.8% were black which compares well with the 5.5% for the whole population in the 1991 Census. When asked indirectly whether mathematics was a barrier, Africans and Caribbeans showed the highest anxiety. When asked directly each sub-group of minority ethnic groups had a higher percentage stating that mathematics was, for them, a high barrier than for the cohort. To understand this situation more clearly, visits were made to two Access courses for Asians in Bradford. The interviews with the Asian students suggested three sources for their higher anxiety in mathematics. Firstly their recollections of school suggest an undercurrent of racism in their teachers. There was a definite perception that teachers had had low expectations of them and so achieving success in mathematics had not been seen by either pupils or teachers as important. There was also a clear picture of family culture and pressures particularly mitigating against Asian girls doing well at school. Some parents had feared the freedom and independence that education could bring and had discouraged study insisting that, for the girls, helping in the home took precedence over study. Finally many of the group had been confused and intimidated by the language of mathematics. This problem, not confined to minority ethnic groups, is exacerbated if English is a second language and the mathematics taught is located in an alien culture.

Does racial differentiation matter?

Certain minority ethnic groups are achieving below their capabilities in initial education. From several perspectives, adult education has an obligation to counterbalance this. First and foremost, adult education has a moral and ethical commitment to equal opportunities based on ideas of equity, access and individual fulfilment. Pragmatically, adult education is funded for its contribution to a well-educated workforce. Legally, the Race Relations Act of 1976 made both direct and indirect discrimination on the grounds of race unlawful, where direct discrimination implies treating a person on racial grounds less favourably than others, and indirect discrimination consists of applying a requirement or condition which intentionally or not has a disproportionally adverse effect on a particular racial group and which cannot be justified on non-racial grounds. In the race relations field, three factors have been identified which have influenced local authority development of equal opportunities: spontaneous protest, pressure for community resources and planned political struggle (Ben Tovin *et al* 1986). Without such concentration on racial issues, the push to widen access to educational opportunities may result in an increase for some groups yet leave other such as blacks untouched.

An exploration of some of the race-related factors in mathematics education

Racism in society

State oppression remains an everyday reality for black people through racist immigration laws, policing policies, the mis-education of black children and the criminalisation of black youth. The Commission for Racial Equality (1990) found overt racist behaviour and beliefs, located in cultural domination with structural institutional racism, hindering access to housing, education employment, justice, political representation and hence power. The education system does little to challenge or even acknowledge this reality in its teaching. Black people are disadvantaged in the education system itself, being the recipients of low expectations with few positive role models amongst the teachers and lecturers. Indeed the education system has been accused of being a largely Eurocentric discriminatory institution which plays a major role in perpetuating existing social norms and inequalities in society (Wilson 1995).

Educational factors

Institutional racism in the education system in general and in the teaching of mathematics in particular can manifest itself in many different forms. The cultural content of the curriculum is often presented as positivist and absolutist, decontextualised, abstract, formal and dehumanised. It frequently erects linguistic and cultural barriers which alienate learners. The pedagogy tends to be based on a written individualistic approach rather than oral, co-operative and creative. Again the result is to normalise certain groups such as white middle-class males and pathologise other cultural or racial groups. This is also apparent in competitive assessment procedures which act as a cultural filter erecting cultural barriers, and biased texts and worksheets which stereotype white and minority ethnic groups. This stereotyping is reinforced by a lack of positive black role models and the unconscious racism amongst teachers leads to unconscious discriminatory behaviour and stereotype expectations (Ernest 1991). Any or all of these forms of institutional racism will reduce educational opportunities and hence life chances. Targeting minority ethnic groups whilst the institution itself remains unchanged is to fail these students.

Staffing

Particularly pernicious is the lack of educational diversity in staffing, lack of a multicultural curriculum and the limited degree of understanding which black students feel that they receive from largely white staff. Many tutors are inexperienced in handling racial and ethnic issues and this can lead to not only low academic expectation through stereotyping but also a high demand on minority ethnic students for personal involvement and contribution as tutors turn to them for advice and assistance (UACE 1990). This problem in staffing was investigated by Leicester who in 1991 sent a questionnaire to mathematics departments in all British universities. The following replies illustrated two diverse reactions.

> I believe some of your questions are not relevant for the teaching of university mathematics . . . What would be the point of putting out information leaflets in other languages?
>
> One cannot say that there is a general awareness, in staff or students generally, of the relevance or importance of an understanding of racism and its roots, and of a non-Eurocentric perspective in a subject such as mathematics.
>
> (Leicester 1993: 100)

Teacher education

The Swann Report (1985) stressed the urgent need for black teachers in Britain's mainstream schools. This can be extended to all branches of the education system. Problems in recruiting to the teaching profession can be partially located in racism in schools and restricted career opportunities. The Commission for Racial Equality (1990) has suggested that racism and racial discrimination is experienced by black students at all levels of the education system. Research on racism experienced by black students on Initial Teacher Education courses does not bode well for their continuance in the system nor for the attitudes which will be imparted by their fellow white students when they start to practise teaching (Siraj-Blatchford 1990). Criticisms by the black students included the ethnocentric and Eurocentric nature of the curriculum and the absence of quality or sufficient discussion of multicultural and anti-racist issues. The courses used discriminatory interview procedures, failed to address anti-racism actively and were not concerned with providing understanding of race equality and power structures. Students experienced racial harassment and discrimination from lecturers and fellow students. Students found that they were expected to act as a knowledge base and catalyst for change but that this was at great personal effort and expense. Even more worrying, school practice provided the worst experience of racist discrimination for many black students. The researchers concluded that policies and practices, both hidden and overt, demanded urgent reform.

Who does it matter to and why?

The five ideologies of education outlined by Ernest (1991) are again useful to the discussion. The most powerful ideology in today's society is that of the industrial trainers whose influence pervades all education including adult education by its control of power through allocation of funds. Their belief that anti-racism in mathematics education is illegitimate is illustrated by the quote at the beginning of this chapter and stems from an even more fundamental belief that mathematics is, in all senses, neutral. They believe that to introduce anti-racist mathematics would represent political intervention aimed at undermining British culture and values. They deny problems in the education provided for minority ethnic groups in Britain, actively opposing anti-racism, promoting a monocultural view of Britain and its history, and continuing to assert that mathematics and mathematics education are value-free and unrelated to social issues.

The old humanists, with their commitment to the abstract and value-free nature of mathematics, see the issue of race as irrelevant to mathematics education. The technical pragmatists, with their utilitarian aims, recognise the waste of talent caused by the underachievement of minority ethnic groups but consequent multicultural initiatives and awareness are directed solely towards employment needs.

The dominant educational ideology of adult educators is that of the progressive educator, being person-centred and concerned to meet the individual needs of all learners. The strategy of this group is to make mathematics available and relevant to learners in a safe supportive environment. The work of Knowles (1970) illustrates this approach. By starting 'where the student is' and using the students' experience as a valuable resource, the adult educator from this tradition will counter prejudice in the classroom, develop the self-esteem and confidence of minority ethnic group students and add a multi-dimensional context to mathematics education through, for example, discussion of alternative number systems.

There is much that is valuable and effective in this strategy, but the culture-bound nature of mathematics is not recognised. The overall framework of institutional and societal racism is not addressed. The emphasis on individual need and development moves away from ideas of collective need. The emphasis on a secure environment halts discussion in the classroom of the contentious and difficult issues of racism. The individualisation of learning locates problems of learning within the individual rather than in society and conceals the nature of power and knowledge. It leads to the introduction of procedures that support minority ethnic group students and are important in the retention of students but that are in themselves unlikely to fundamentally change the composition of adult education. This approach is concerned and responsive, and operates on the assumption that the learner is empowered by education and can then choose, if they wish, to attempt to change society. It stresses education as an agent of personal change rather than social, political and economic transformation.

In contrast, the final group in education – termed 'radicals' in adult education and 'public educators' in compulsory education – locates education within its political, social and economic framework and hence sees the under-achievement and under-representation of minority ethnic groups in mathematics education as a result not of individual failure or racial inferiority, but of structural and institutional racism. Radical educators work at three levels – explicit discussion and consciousness-raising, the selection of the content of the mathematics curriculum and modes of pedagogy and organisation employed.

However, the industrial trainers at present control finances and hence the system. The educational ideology of many adult educators may be that of the progressive educator with perhaps a radical tinge. Moreover, the individual adult educator may have a belief in the ethos and values of equal opportunities and even a strong commitment to social change. However, these items may be forced into the background of their mental map by more urgent imperatives which stem from the Government's, and hence the funding bodies', vision of education for social conformity and a well-trained workforce.

What can be done to overcome race differentiation in learning mathematics?

The barriers to participation are located in policy, what is taught, how it is taught and who it is taught by. For both increased and effective minority ethnic participation, it is useful to remember that 'there are no logical barriers to open learning but rather material and cultural ones' (Griffin 1983: 86) and it is these barriers which need to be overcome in order for the minority ethnic groups to participate and to succeed.

Policy, targeting and monitoring

The Swann Report, *Education for All* (1985), although concerned with the education of minority ethnic group children, is also relevant for adults. The central issue as identified by Swann is the development of a multicultural education and the recognition that the problem is not how to educate the minority ethnic groups but how to educate everyone within a multi-racial and multi-cultural society (Taylor 1990). A crucial move is the establishment within the institution of an equal opportunities policy expressed not in vague and general terms but which clearly draws out specific issues. One such should be ethnic diversity. Within the framework of such a policy, institutions could increase minority ethnic group participation within a two-pronged approach of targeting and monitoring policies and appropriate course provision.

At least one branch of continuing education stands accused of being characterised by inertia in the area of equal opportunities. A Universities Association for Continuing Education survey of university departments of continuing education showed little evidence of multi-cultural developments or clear policy guidelines (Taylor 1990). The first step to counteract such inaction would be the adoption of an anti-racist policy statement for the institution. This might include statements of principle, guidance for action and monitoring processes. These would then provide the framework for provision and inform decisions concerning recruitment of students, needs analysis and outreach, publicity, selection procedures, prior qualifications and experiences, curriculum change, access routes, collaborative systems, access to assessment, physical access, student support, financial access, planning and implementation, guidance, course structure, funding, marketing and staff recruitment and development (NIACE 1989). In short, the policy statement would inform the teaching and learning process and the policies, structures and practices of the institution.

However, it is crucial that cultural pluralism should not be tokenistic, where other cultures are understood superficially or through the assumptions, values and beliefs of dominant groups. Institutions need to move away from the normalisation of certain cultures and the pathologisation of others. What is required is pluralism and diversity at a conceptual level within institutional planning and policy where the perspectives and values are anti-racist and counter the cultural impact of racist non-formal learning. Within the parameters of the policy statement, course planners need to be aware of the social and economic realities faced by black people and work jointly with local communities. Issues of accreditation, advice and guidance and student support are important. Positive action strategies may be required to attract black people onto courses and black staff into employment in non-marginal posts. The curriculum needs to incorporate equality perspectives, black interests and anti-oppressive teaching materials together with assessment procedures that take account of the experience of black students. Black people will be needed on advisory and decision-making committees. Awareness will need to be encouraged through policies, staff development programmes, marketing and publicity strategies and recruitment policies. Substantial institutional changes may be needed.

For minority ethnic group participation to increase, it is arguable that commitments to equal opportunities and widening access are not enough in themselves. There need to be active recruiting, marketing and targeting policies, with agreed target figures and methods of implementation and evaluation. Without targets, little will change. They provide the incentive and rationale for positive action strategies. Programmes need to re-examine their selection criteria, particularly when there are more applicants than places. In line with the targeting policies of the programme, it may be appropriate to introduce positive discrimination or, at the very least, criteria that are located in the targeted groups rather than just motivation and commitment, important though these are. Motivation and commitment are nebulous qualities which can be shaped by the interviewer's cultural norms and expectations.

Targeting needs to be focused more specifically than the broad categories of white and black. But for targeting to have any real meaning, there needs to be an annual or periodic review of policy objectives with a view to effective implementation and the systematic collection of statistical data relating to student characteristics. Such data is crucial for effective monitoring which relates outcomes to targets. The kind of information monitored could include whether enrolments are *pro rata* with representation in the local community, the uptake of different courses by different ethnic groups, completion and other performance outcomes, staff recruitment and structuring. Ethnic monitoring is a complex and sensitive business. It must be based on self-classification and confidentiality, and must be voluntary.

There are examples of good practice which derive from an institutional commitment and lead the way in attracting ethnic minorities onto adult education provision. Work at the Centre for Access and Advice, University of East London has shown how outreach work is particularly important in accessing minority ethnic communities. Benn and Burton (forthcoming) found that outreach work is essential if the institution wants to reach *all* its targeted groups, as it allows information to be spread through a network of local groups and individuals. Initially a full-time outreach worker, preferably someone to whom the targeted community can relate, needs to spend time in the community networking, creating links and establishing trust and confidence before courses can be developed. Each area will be different and each area will have different problems. Inner cities tend to have diverse populations, even when crude analysis may suggest that only one or two different ethnic groups live there. The reality is often somewhat different with many different groups co-existing side by side. In many cases, factors such as religion are as important as race. Advertising which is appropriate to all targeted groups is crucial. The Centre uses leaflets in seven different languages and runs access 'fairs' that involve the whole community but are directed particularly at the targeted group. Some fairs need to be exclusive in order to attract their target group (for example, Muslim women). Good educational counseling and guidance is essential, not located in the institution but through the outreach worker in the community. East London has found that once this model is up and running, new clients are generated through pro-active role models (people who have been successful in completing courses) and that their outreach work has proved effective in reaching groups 'blocked' by cultural and/or material barriers.

In another innovative scheme, Bradford and Ilkley Community College, in partnership with Bradford City Council and local schools, offers young Asian men and women who have been denied opportunities through the normal academic routes a way into HE and FE which circumnavigates some of the traditional barriers. This partnership has been set up as part of Bradford City Council's positive action policy which both addresses the need for alternative routes into HE/FE for people from minority ethnic groups but also provides a pool of much needed bilingual teachers. The project offers students a two-year paid classroom assistantship whilst they study part-time on an Access course. The aim of this project is to encourage more people from minority ethnic groups onto the four-year BEd programme and then into teaching.

Curriculum

Mezirow's work on perspective transformation and fostering critical reflection is useful to this discussion (1977, 1981, 1990). He argues that many, if not most, children leave school with deeply racist assumptions and that adulthood is a time for reassessing this early learning. He defines meaning perspectives as the structure of assumptions that constitutes a frame of reference for interpreting the meaning of an experience. Critical reflection is the assessment of the validity of the presuppositions of one's meaning perspectives and the examination of their sources and consequences. Through critical discourse, learners develop their meaning perspective to allow a more inclusive, discriminatory and integrative understanding of experience. The adult educator's task is hence to help the learner understand and change the meaning perspectives and distorted views of reality formed by (perhaps racist) societal and institutional ideologies and internalised by individuals. Mezirow's ideas of perspective transformation show how a disorienting dilemma leads individuals to reflect critically on their own and society's meaning perspectives and in so doing change their own meaning perspectives in order to be able to assimilate the new experience. The disorientating dilemma may arise from an encounter with different cultures that challenge early prejudices. This approach encourages the adult educator to confront the learner with alternative world views and paradigms. This is change at an individual not societal level but the raised awareness forms an infrastructure for social action if the individual so chooses. The critical reflection of the student can lead to transformative action and hence social change.

This critical reflection can be an integral part of the adult educator's mathematics curriculum. This could start with a discussion of the socially constructed nature of mathematics knowledge and hence its fallibility, the multicultural origins of mathematics and the validity of the mathematics of all cultures. The racist nature of society and institutions could be illustrated and condemned, the social context of all education discussed and the relationships between power and knowledge exposed. This would lay the groundwork for an analysis of resultant problems encountered by certain groups in society. This could lead to an exploration of the position of mathematics and statistics in our society (for example, the social statistics concerning minority ethnic groups in Britain and globally) and the role of mathematics in controversial and value-based issues such as government statistics. Multi-cultural mathematics, ethnomathematics and the ethnocentrism which occurs from the commission of use of racist materials whilst omitting a black perspective could be discussed.

Teaching based on passive didactic modes of learning are unlikely to get learners to examine or change their own prejudices, assumptions and stereotypes. Learning is most effective when based on active participation through discussion, based on experiences and reflection. This allows the learner to draw on their own experiences as real knowledge. Differences in experiences and backgrounds allows different voices to be heard in the learning conversation. The pedagogy could be based on a variety of learning approaches but involve co-operative projects, problem-solving and discussion applied to the thorough and rigorous treatment of mathematics required for external assessment. The teacher's role is the difficult one of respect for the learner whilst acknowledging the power asymmetries of all formal learning. Materials need to be screened for bias and stereotyping and where this occurs, it can be discussed and made part of the educational exercise.

Participatory democracy is fundamentally pluralistic and entails the acceptance of the intrinsic worth of all human beings and their unique individuality. This means that cultural differences should be viewed as differences not deficits (Pai 1990). Adult education provision already concentrates on the common problems of the adult returner often with little or nothing in the way of formal qualifications. But, in addition, the course content needs to bring the concept of structural inequality to the surface by incorporating explicit anti-racist teaching in the curriculum by open discussion of issues of race, gender and class. All cultures can be valued by the use of a diversity of materials which is anti-racist and/or pluralist. Whether the curriculum places emphasis on individual educational development or political education through social action is at the discretion of the provider but what to do with the knowledge acquired is at the discretion of the learner.

There is a need to incorporate exemplars from cultures and traditions that have been ignored or devalued for too long to support the principle that learning should be tailored to social and physical environment in which the learners' live. Utilising the rangoli patterns which decorate the homes of Hindu and Sikh families, the geometric art which forms the basis of the Islamic designs in mosques and wall coverings, and the calendars which determine the Jewish and Chinese New Years would draw on a rich heritage whilst recognising and valuing other cultures. This approach would also introduce a more holistic approach by linking mathematics to art, design, history and social studies.

Emblen's (1991) work with children of Asian families in English schools shows that what is required are teaching schemes that focus on strengths not deficiencies and take into account issues such as English not being the first or home language. Learners should not be confronted with unfamiliar ideas and unfamiliar language at the same time. This means that the mathematics curriculum should provide opportunities for developing language as well as mathematical skills. It requires that learners are not assessed by something other than mathematics and in which they are not fully competent: the English language. It also means, when appropriate, accepting the meaning and intention of a communication without criticising the form it takes. Everyone should be allowed to formulate mathematics ideas in a familiar language.

Anderson (1990) and Joseph (1987) both show that the present structure of mathematics education is inherently Eurocentric and consequently the curriculum 'reinforces the racial and sexual inferiority complexes among people of colour and women' (Anderson 1990: 349). Anderson argues that mathematics teachers have to break down the walls of Eurocentrism and place mathematics within its historical and cultural setting to 'shatter the myth that mathematics . . . is a "white man's thing" and [to emphasise] . . . that all civilisations . . . are bound inextricably to each other' (1990: 355). He has developed a non-Eurocentric approach to mathematics teaching that endeavours to ensure the relevance of mathematics to his mainly black students. The early sessions are discussions on the historical, cultural, and socio-political implications of mathematics. At all stages an emphasis is placed on the role of other races and cultures in the development of the subject. The importance of mathematics to real people in real life is drawn out by regular class discussions of current issues in the social and natural sciences, the development of technology and job market skills. The emphasis is on the quality of mathematics knowledge rather than the quantity, thus reducing the time pressure. The creation of study groups allows collective study and in-class group work. Anderson claims that this approach leads to students having a more positive, self-assured attitude about themselves successfully doing mathematics with the consequence that failure and drop-out rates fall considerably.

Provision for non-traditional users is most accessible when offered, at least initially, within communities. Distance learning can also be harnessed by the use of not the latest and most expensive technology developments but simple technologies already in people's homes. Video recorders have long been staple items in many minority ethnic homes because of the usage of mother-tongue videos. This has already been recognised and utilised by a number of providers. Learning based in the home can assist in intergenerational patterns of learning.

The problems here, as in other forms of equal opportunity provision, are many, including the potential clash between instrumental goals and social goals. There may be conflict between this approach and previous school culture and/or other cultural backgrounds and with other ideologies of education. There is the persistent dichotomy between multicultural and anti-racist education, between education for personal and individual development or for social change. There is the problem of stressing ethnicity and the extent to which this undermines issues of gender or social class. The path to the equal opportunity curriculum is not a straightforward one nor an easy option. That is not, however, an excuse for lack of action.

Conclusion

There is an on-going debate between the proponents of multi-culturalism and those of anti-racism. A liberal multi-cultural approach is through gradual partial improvement but this can be criticised as being tokenist and irrelevant to any radical innovation or change. An anti-racist approach requires a more radical change within our institutions and society itself which will be fought bitterly by groups with a vested interest in the *status quo*. This dilemma will not go away but should not be allowed to hinder moves through either approach towards the ultimate goal of *mathematics for all.*

Chapter Fifteen

Mathematics, work and play

Mathematics as work

The problem with learning mathematics can be linked to the perceptions of mathematics in its cultural form – a combination of mystery and power. For many, including or indeed especially mathematicians, mathematics is quintessentially pure reason and beauty. It exists unchanged for all time. It is written in the stars. This ensures a mathematics that is exclusive, elitist and disconnected from everyday experiences. The consequent abstraction and alien discourse reduces mathematics to a gift from tutor to learner, a peek into the mind of God, which many see through a glass darkly, if at all. This fundamental unfriendliness of the epistemology ensures that for many doing or learning mathematics always feels like work, never like play.

Mathematics for pleasure

This is unfortunate because there is a need for humour, laughter and play to dissipate tensions that block learning and allow people to tackle sophisticated skills and concepts which presented in too formal a way would seem too difficult. One of the areas of mathematics which causes learners great problems is statistics. Darrell Huff's wonderful book *How to Lie with Statistics* (1985) is an excellent example of conveying concepts with humour. Two extracts illustrate. One of his many cartoons has one character recommending to another over drinks, 'Don't be a novelist – be a statistician, much more scope for the imagination . . . '. Another example that I quote time and time again in my teaching is in the chapter 'Post Hoc Rides Again'.

> A positive correlation has been found to exist between the number of children born into a Dutch family and the number of storks' nests on the roof of each house. The proof of an ancient myth or the fact that big houses both attract large families and also have more chimney pots on which storks may nest?

Another personal favourite of mine is Isaac Asimov's *Asimov on Numbers* (1977). In an interesting and humorous way, he covers topics such as the history of numbers, the importance of the number zero and the reality of imaginary numbers. These and many other authors show that mathematics needs to be taken seriously but not solemnly. This critique is not to advocate 'easier' or less demanding learning. Research has shown that courses which set out to be fun are enjoyed *because* the work is hard (Rodgers 1990). Successful learning in formal setting requires relaxed, but high, expectations of success, lots of encouragement, believable validation, respect, sufficient challenge linked to sufficient experience of success. Humour helps to encourage the supportive environment in which this can take place.

Ernest (1986) claims that games can teach mathematics effectively by providing reinforcement, practical skills and motivation and hence helping the acquisition and development of concepts and problem-solving strategies. Mainly (1991) extends this by asserting that games not only enable learners to *learn* mathematics but also to *do* mathematics. An example of a simple but fun mathematical game is Fizz-Buzz, a verbal game based on counting and multiplication. There are many others which adults play for pleasure at social gatherings. Playing games *is* doing real mathematics if real mathematics is that which is important and meaningful to the learner, using mathematical processes and thinking in a mathematical way – but this may need to be made explicit to the learner.

In the learning environment, a meta-language is needed to enable the tutor and learners to discuss the mathematics used in games and the processes of learning in this way. People use reading for pleasure, to get information and to help them understand and cope with their lives. To use mathematics in the same way, it needs to be important and useful, independently of the formal learning situation. Games are one way of providing this. People often use mathematics to help them predict the outcome of events (if I buy this new washing machine, will I go overdrawn in the bank?). Games allow prediction in an important but not dangerous environment. They allow judgements as to 'best' moves (conjecturing) hence developing relational understanding rather than rote-learning. Learners may begin to devise generalised strategies for particular games, an important process in real mathematics. They provide opportunities for checking (one's self and one's opponents), justifying and also discussion, co-operation and the development of group dynamic.

The phenomenal sales of puzzle books illustrate the wide enjoyment gained from puzzles. Puzzles are very different from textbook mathematics problems. The latter are always designed to contain only and all essential information and usually point out, either implicitly or often explicitly, the technique they wish you to use (for example, 'use the substitution rule to . . . hence or otherwise . . .'). Puzzles are more like real-life problems. They are full of redundant information, often missing necessary data. There is rarely one answer and no one cares about the method. What is required is an effective solution preferably using a short cut. In addition, puzzles often deliberately try to mislead. Eastway (1995) illustrates that the difficulties many people have with mathematics are often located in terminology as in the following example.

> A particle travels in a straight line through one medium at speed 3v, and through a second medium at speed v. What is the path of

> the particle such that the time it takes to get from point A in the first medium to point B in the second medium is minimised?

He contrasts this with the different wording of a problem called Baywatch which has the same method of solution.

> Wayne is a lifesaver and sees a swimmer in distress. He wants to get to the swimmer as fast as possible, not least because he is being watched by 20 adoring Californian babes. He can run faster than he can swim. Should he aim straight for the swimmer (the shortest distance) or run to the part of the beach nearest the swimmer so that his swimming distance is at a minimum?

The puzzle interests the reader, contains superfluous information, misses out necessary information and points to two alternatives, neither of which is correct. It is no contest which most people would rather solve!

Similar ideas were used by Yorkshire Television's programme *Fun and Games* which ran for several successful series in the late 1980s. The rationale behind the programme was to capitalise on people's interest in puzzles as a vehicle to think about the embedded mathematics concept (Hoyles 1990). It incorporated mathematical ideas ranging from recursion and probability to permutations, topology and four-dimensional space. The mathematics was made explicit by a mathematician (Celia Hoyles) whose role was to try to pinpoint what was difficult in the problem, to introduce mathematical ideas, work with them a little in the discourse of mathematics and finally apply what she had done to the puzzle. Whilst acknowledging that mathematics is much more than puzzles, she argues that this approach helps not only with the learning of mathematics but also with improving the perception of mathematics in people's lives.

There are other occasions where mathematical thinking is used as part of a pleasurable activity. Harris (1995), for example, found that the range of mathematics included in courses on knitting, curtain-and-blind making and machine patchwork included arithmetic, geometric, algebraic and problem-solving skills. She also identified that all the on-task conversation of participants was mathematical. However, though the women participants demonstrated a conceptual understanding of a range of mathematics equivalent to 'basic' levels and more, they lacked self-confidence in their mathematical ability born of a combination of lack of both recognition and the language and symbols of mathematics. Similarly, Black's work on knitting and mathematics shows the considerable mathematics content of this pleasurable activity and suggests that drawing out the implicit mathematics within knitting empowers the knitter as both knitter and designer but also through increased confidence as an individual (1995).

Many of these pleasurable processes are not thought of as mathematical. Indeed it sometimes seems that if people can do it, it is called common sense; if they can't, it is seen as mathematics. Empowering learning needs the stages of reflection to allow the participant to understand the 'real' mathematics that they are doing with the consequent increase in self-confidence.

'Really useful' mathematics, or mathematics that works

Many mathematical processes, pleasurable or otherwise, are not considered as mathematics but just what needs to be done to solve a problem that has arisen. We now examine mathematics constructed by individuals or social groups to cope with a particular work environment. Carraher (1991), surveying the extensive research in Brazil into the use of mathematical concepts and procedures by people in low-paid jobs, found that though there are indeed differences between academic mathematics and work-based mathematics, there are substantial conceptual and empirical overlaps and the practices, representations and procedures in formal mathematics influence informal mathematical approaches and *vice versa*. Formal mathematical knowledge arises collectively usually constructed by academics, is then recontextualised by, for example, textbook writers and then transmitted to learners by teachers (Dowling 1991). Informal mathematics is created by individuals or social groups and is then transmitted informally through, for example, the work situation (sitting next to Nellie). It is learned in the course of a social activity such as work.

The Brazilian studies showed several interesting findings. For school 'failures', the success rate in solving problems was very high in a real-life work setting compared with the success rate in written problems even when expressed in work language. Procedures seemed qualitatively different from those taught in school. However, using Vergnaud's theory of concepts (1983, 1989), Carraher argues that formal taught written mathematics routines such as column multiplication or 'borrow and give back' subtraction are based upon the same properties as oral informal mathematics. The common features are certain invariants of mathematics such as the additive composition of quantities. What differs is their symbolic representation. This leads to the argument that what is valued in our society is not the ability to successfully perform the invariants of multiplication but the ability to utilise present-day symbolic representation *ie*, column multiplication. Those who are successful in informal mathematics but not formal mathematics demonstrate knowledge of the fundamental invariants but have difficulty with the particular procedures and representation adopted by formal mathematics. Informal mathematics shares the invariants of formal mathematics but differs in the symbolic representation: all mathematics have a common set of invariants but a different discourse. This strengthens the suggestion made in Chapter 10 that study is undertaken to enable the learner to move from the discourse of their own ethnomathematics to the discourse of formal mathematics.

Carraher (1991) found that in work situations, workers usually gave answers which were correct or 'sensible'. Students asked to complete the same computation generated a wider variety of strategies but were more likely to lose the meaning of the problem. They often gave nonsensical answers. Whilst acknowledging that self-discovery will take people only so far in the development of mathematical knowledge, Carraher concludes that work provides challenges and opportunities for people to develop mathematical knowledge which is meaningful. This is supported by two other research projects on work-based activity.

Scribner's work (1984) in a factory environment showed that in a work-based context, the mathematical problem-solving of the unskilled workers in their own area of work was more efficient, more appropriate and more sensible than that of either a group of students or a group of skilled workers who shared a common cultural knowledge of the work but who did not actually deal with the particular problem or products themselves. The skilled workers and the particularly the students tried to find an algorithm to solve the problem with little attempt to adapt this to the problem in hand. This mode of solution, whilst very powerful if correct, is vulnerable to an incomplete grasp of concepts.

Lave's work (1988) with 'experienced' supermarket shoppers showed that the success of arithmetic problem-solving in a test and shopping situation was quite different. The shoppers' scores averaged 59 per cent on the arithmetic test compared with an incredible 98 per cent in the supermarket. This demonstrates the high level of success where people develop their own solutions to problems located within their own social contexts. This mathematical knowledge can either be developed by the individual or handed down through a community. The major advantage of this mathematics is its meaningfulness, but it is limited in the conditions to which this knowledge may be usefully applied. Formal mathematics on the other hand seems to fail many people, but be more useful to those who have had a longer period of initial education. It takes time to understand the discourse of formal mathematics and to develop the ability to recognise and deploy the symbolic representations. The early stages of this process are characterised by a total inability to handle these concepts through to use of inappropriate strategies and nonsensical solutions. However, the eventual payoff is a set of powerful methods which accomplish a lot of work in a little time and with little effort.

Differences between 'mathematics that works' and academic mathematics

Different contexts will affect the rightness of answers but also the methods and language used. Although the invariants may remain, the symbolic representations and procedures will vary. Where the problem is located within the individual's own discourse or ethnomathematics, the solving strategies are likely to give 'correct' answers. Where formal mathematics is used, the smaller number of 'correct' answers is probably due to a combination of forgetting the concepts, rules and strategies, partially remembering methods, and unfamiliarity with the context.

Even where rules and methods are remembered, it is not necessarily easy to transfer from one mathematical discourse to another, for example from work mathematics to formal mathematics or indeed, *vice versa*, because transfer involves major differences in discourse. Each discourse is governed by language, social and power relations, resources and many other factors. All these may affect the ability to transfer knowledge. Just as there may be severe emotional blocks in formal mathematics, so also may power and discipline issues affect informal mathematics. These may inhibit effective performance. An example of this may be seen in the gendering of mathematics where women's position in society and in the workplace affects their ways of thinking about mathematics. This might result in very negative feelings about real life or workplace mathematics with consequent emotional blocks in this mathematics which is as strong as the mathsphobia of formal mathematics.

Formal and informal mathematics may range over the same topics but the how and why are very different. There are major differences located in content and context. The social group basis of informal mathematics consists of family, work colleagues and so on, whereas that of formal mathematics is through external bodies such as society or the government. Informal learning links intellectual and emotional factors whereas formal learning ignores the emotional. Methods for informal learning include imitation, identification, co-operation and observation whilst formal education relies heavily on language, instruction and competition. In formal mathematics, the solution is an end in itself. In informal mathematics, the solution is often just part of a wider problem; it is just a way with dealing with part of one's world that needs mathematical thinking (Scribner and Cole 1973; Gerdes 1996). Many informal problems start with asking 'What is the problem?' rather than the command 'Solve the problem!'. In everyday life, mathematical problems are not a matter of 'can do it' against 'can't do it', but a variety of possibilities to be drawn on. It is only when asked to write out the method, that people forget these choices.

This can be summarised in Table 3.

Table 4 illustrates this difference.

Theories-in-use

Schon's concept of theories-in-use, taken from his reflective practitioner approach, may help the understanding of the dichotomy between formal and informal learning and the impact this has on the individual attempting to do or learn mathematics (Bright 1992). Theories-in-use refers to implicit, informal and incidental information and knowledge which guides and assists the design of action within an ongoing and dynamic situation. The concept resides at an informal, implicit level and can be applied to all types of human action and involves the difficult notion of hidden, implicit and tacit theories underlying action. In order to engage in any mathematical action, an individual must bring into play such a theory-in-use.

Theories-in-use are at three different levels. The first is conscious and

Table 3: A comparison between formal and informal mathematics.

Informal mathematics	*Formal mathematics*
embedded in task	decontextualised
motivation is functional	motivation is intrinsic
objects of objectivity are concrete	objects of activity are abstract
processes are not explicit	processes are named and studied
data is ill-defined and noisy	data is well-defined and tidy
tasks are particularised	tasks are generalised
accuracy is defined by situation	accuracy is assumed or given
numbers are messy	numbers are designed to work out well
work is collaborative, social	work is individualistic
correctness is negotiable	answers are right or wrong
language is imprecise and differentiated	language is precise and carefully responsive
	(Harris and Evans 1991: 129)

Table 4: An illustration of the different approaches in formal and informal mathematics.
A car park charges £1.00 for the first hour and 25p for each additional hour or fraction of an hour. For a car parked from 10.45 in the morning until 3.05 in the afternoon, how much money should be charged?

Informal	*Formal*
mental arithmetic used	pencil and paper
data gathered from watch, ticket, attendant, board	complete data provided
round figures to ease calculations	data provided is exact
use 'easiest' mental algorithm	use standard algorithm
estimate cost	exact correct answer
ask for a second opinion	work in isolation
work till reasonable answer obtained	work in given time
use previous experience to judge 'reasonableness'	do not utilise previous experience
use answer to check have enough to pay	no further use for the answer
	(Hind 1993: 16)

public and as such only contains information, opinion, beliefs and attitudes that are widely acceptable. This can be seen to reflect academic formal mathematics. The second is conscious but private, not for public airing. This can be totally private or shared with selected others, and can be seen to reflect the individual's ethnomathematics when it is known at a conscious level. The third level is unconscious which reflects the individual's ethnomathematics when the individual does not realise in a conscious way that they have such a mathematics. Argyris and Schon (1974) argue that contradictions between the espoused (public) theories and actual (private, conscious or unconscious) theories are the most important source of failure to engage in effective learning. This illustrates the need in mathematics teaching to encourage the learner to identify and make public in a validating environment their own ethnomathematics. The contradictions between the different mathematics or theories-in-use can then be recognised and more easily dealt with. Even understanding the existence of the different mathematics and the problems in moving between them can be helpful in the learning process.

The different methodologies for solving a simple mathematical problem can be illustrated with subtraction. Gerdes (1996) found that when asked to find 62–5, three main methods were used.

59% calculated 62–2=60; 60–3=57.

29% calculated 60–5=55; 55+2=57.

12% calculated 62–10=52; 52+5=57.

These diverse approaches to solving such a simple problem suggest a wide range of private mathematics in any adult mathematics classroom.

The predominance of academic mathematics – the usual conscious and public mathematics – is due to cultural socialisation and reflects powerful, external pressures. However, learning this mathematics is often solely instrumental, technical or tactical within a given set of goals and objectives which are not reflected upon or questioned. How can they be when they are so firmly endorsed by society? More effective learning involves critical reflection on action and continuous redefinition of goals and objectives where knowledge is seen as tentative, open to question, crucially dependent on evidence and changing, and which cannot be applied out of context without amendment or abandonment. This approach cannot be applied where goals, objectives and knowledge generation is owned by others. Effective and competent action in mathematics, (mathematics which works) requires effective and competent information and knowledge-generation processes. It requires a clear understanding of the different mathematics in life and the skills to harness the most appropriate to given situations.

One way of developing this form of learning is through the use of open-ended questions (Spencer 1996). A problem posed in class is open-ended if there is more than one way to solve the problem; there is more than one answer to the problem; or the problem requires the student to interpret the question or make a value judgement. Taking my current real-life decorating problem – a closed questions would ask:

> if a roll of wallpaper costs £7.99 a roll, how much will it cost to decorate a room which needs 6 rolls?

This can be opened up by saying, 'I have £100 to spend on decorating a room; how would you do it?' This approach of starting from the answer can be applied to a wide range of problems and to various ability levels. By working in groups, learners develop co-operative skills but also learn to reason and communicate mathematically. Relevance and links to each learner's previous experience of mathematics and culture will arise in the discussion, facilitating meaningful multicultural mathematical learning and empowerment.

Work means 'bad', play means 'good'

Walkerdine (1988) discusses a further interesting interplay between mathematics and the notions of work and play. She links the distinctions between work and play with those between rote-learning/rule following and 'real' understanding, arguing that work is the basis of the 'old' discourse and that in the 'new', learning is founded on activity and play. Play has become a fundamental device for mathematics learning. This concept of play is located within the concept of learning as the joy of discovery and pleasure in ideas. Constructive play develops experience and discussion out of which mathematical structures, concepts and strategies (mathematical learning) will emerge. An essential ingredient of this approach is the need for individuals to develop at their own rate.

This 'new' discourse leads to the concept of student-centred learning through experience and activity. It is a reaction against 'work', where work is teaching didactically to students in rows. Work reflects a 'banking' concept of education whereas play reflects the 'liberating' education acquired through active exploration and utilisation of the learner's environment. Work is symbolised as rote-learning and rule-following. Work is imposed and regulated, play is student-generated, fun and freedom or liberation. Real understanding, with its associated possession of power and control, grows out of play; only rote-learning, with its associated notions of powerlessness and being controlled, grows out of work. Society only values attainment achieved through 'real' understanding. One of the results of this is that certain ways of learning are esteemed, others are not. Chapter 11 illustrates this by suggesting that this is the reason that girls' achievements are so often denigrated as 'just due to hard work'.

This concept of 'work is bad, play is good' has permeated the adult classroom, substantially altering the dynamics of the classroom by validating student-centred learning through experience and activity. Whilst recognising much of value in this approach to helping adults to learn mathematics, Walkerdine's insights help us to recognise the dichotomies within this paradigm that are hidden by the cosiness. Power and control are still there in a hidden but just as powerful form which results in an illusion of power. Choice and control are still limited and certain voices are still heard more clearly than others.

Conclusion

In this chapter we have raised various links between mathematics, work and play. Some are pedagogic. The increased use of humour in the teaching of mathematics, the incorporation of pleasurable activities such as games and puzzles and the drawing-out of mathematical activity from occupations such as knitting, all lie within the power of the adult educator. So also does the responsibility to raise awareness in the learner of the alternative forms of mathematics so that the learner can more easily understand the sometimes contradictory theories-in-use that can cause so many learning and doing difficulties and hence choose for each situation the mathematics that works.

The responsibility to value *all* forms of learning rather than the arbitrary categorisation of some learning as 'real' which contributes to the marginalisation of some groups in society is part tutor's, part society's. Some issues raised are highly political. The disjuncture between formal and informal mathematics seems greatest in low-level work skills and the activities of most people in everyday life. It has not been observed to anywhere near the same extent between academic mathematics and the high level professions such as engineering and accountancy. It is arguable that the education system is designed to produce a professional class who can calculate and reason scientifically and the rest who are reasonable in order to be governed. The informal ethnomathematics of the various groups in society exist within society alongside the professional skills of, for example, the academy, accountancy and engineering.

The choice of which knowledge should be canonised in the curriculum and how this curricular knowledge relates back to society is a political decision which in a democratic society should seek to provide for the needs of all citizens. This again comes down to the intensely political issue of power. If the principle role of the education system is to maintain the social system and preserve the *status quo* with regard to mathematical knowledge, this will work directly against a mathematics that truly incorporates the needs, values and context of all citizens. This illustrates the conflicting goals of the system. If mathematics is used as a means of selection and control, the processes involved will work against enjoyment, involvement and appropriation of mathematical ideas (Hoyles 1990). It will also work against the full utilisation of the abilities of all members of society.

Academic mathematics does have a role to play but as Henderson (1990: 118) says,

> . . . formal symbolic expressions are often excellent ways of capturing certain aspects of our experience. The problem is that formal mathematics has been masquerading as the highest, truest, ultimate form of mathematics. This limits the understandings which we construct of our human experience. This masquerade damages the

> human spirit. Everyday, persons all over the world are rejecting their own experiences because their own experiences do not seem to fit into the formal frameworks which masquerade as the only true mathematics.

Mathematics that is taught with humour and enjoyment; that is seen as intrinsic to many of the enjoyable things in life; that recognises, validates and builds on the skills individuals have developed and learned their own social environment: this mathematics need not be classified as work or play but rather as mathematics for everyone.

Section 4

Implications for practice

Chapter Sixteen

Towards an empowering curriculum

Many adults have a problem with mathematics as was shown in Chapter 8. All sorts of people at every level of academic ability from ABE to graduates are affected socially, politically and economically, illustrating that this problem is not located in the individual but in society. Chapter 7 suggests that three roles of education in our society are to govern the production and distribution of knowledge in society; to act as an agent of socialisation of the individual into the totality of relationships in society; and to be a force for emancipation and growth, both individual and collective. The curriculum to some extent (differing at differing times) implements the 'treble role'.

Until fairly recently there was a consensus that society's future lay in the learning capacity of the young, but more recently the importance of adult learning has gained ascendancy. The reasons for this include a growing urgency at an official level for the flexibility, adaptability and new skills in the adult population which, it is hoped, will turn around or at least slow down the economic decline. In addition, there is a growing cult of individualisation with increasing differentiation of individual identities and a growing preoccupation with personal autonomy and the need to maintain social stability through the incorporation of all cultural groups such as the working class, women and minority ethnic groups.

These developments are not necessarily connected with emancipation. Indeed, the crisis in work has tended to foster autocracy and allowed the dominant forces in society to seek to ensure that the most developed part of the adult education system is that which is designed solely to upskill the workforce. Both mass citizenship and individualisation have contradictory implications for emancipatory education, arguably resulting in increased personal autonomy with the potential for emancipation but also for incorporation. As a result, the emancipatory role of adult education has been substantially eclipsed by its socialising and distributive roles.

Further, the present condition affects learners and tutors as well as the curriculum. Writers such as Field (1992) have questioned whether adults are losing their capacity for other than technical or cognitive learning in a society where they feel so powerless and insecure. At the same time, tutors who have reform-orientated values which they try to live out in their working life through their pedagogy and curriculum development are employed by the state. If their aspirations are at odds with the perceived values and practice of the state, then this may cause unease in their position of receiving state money whilst subverting state intentions.

Nevertheless, to retain genuinely its diversity of function, education should contribute to emancipatory learning as well as socialisation and distribution. If this is to be the case, part of the task of the curriculum is to enable learners to acquire knowledge and skills in mathematics in the hope that this will allow them to acquire better control over their destiny whilst giving tutors 'permission' to facilitate this. This implies a curriculum which creates and practises democracy, giving free space for choice and communication, within the education setting. This requires a reconceptualisation of the curriculum by politicians, administrators, academics, tutors and learners within an overall political, social and economic framework. It implies a curriculum which seeks signs of a positive common, as well as individual, future.

Democracy, citizenship and empowerment

An emancipatory curriculum emphasises democratic citizenship as well as the liberal promises of social mobility or equality of opportunity. This implies the learner not only developing the skills to generate and solve their own mathematical problems but also understanding why and how other pervasive mathematics problems are generated and maintained along with their consequences for democracy and citizenship. The curriculum should allow the learner to have as much choice as possible over the direction and content of their learning, encouraging critical reflective learning in a social context through discussion, multiple perspectives and constructive argument. A curriculum premised on democracy and citizenship, as well as training for work, must include an understanding of the significance of numbers within the social, political and economic framework that is our society, enabling the learner to acquire a greater understanding of social issues and actively engage in the process of critical citizenship.

A central feature of the discourse of adult education has been a concern with the concept of empowerment. However, there is a real dichotomy between the essentially liberal humanistic perspective which declares that empowerment starts from within the person and is primarily about personal change, and a more critical and structural analysis which recognises the limitations and wishful thinking of a personal empowerment that does not relate to wider community and societal inequalities. Claims have been made for adult education in raising consciousness, in combating social inequality and in developing collective empowerment when the reality has often fallen short of this. The danger is that adult education is only liberating and empowering at a very individual and restricted level and that it impacts in a very marginal way on local communities, society or the formal education system. This constriction is made more acute as adult education services are unlikely to serve well, or even attract in the first place, those individuals most vulnerable and least powerful in society.

Culture

Whilst acknowledging the continuing power and influence of the absolutist view, the way towards a more egalitarian society lies in a curriculum based on the notions of the social constructivists. The 'grand narrative' of Western academic mathematics needs to respect and allow space to alternative local narratives. The strongest case against absolutist, hegemonic mathematics education is that it has been shown time and time again to be useless and meaningless to the majority of learners.

The introduction of the notion of culture as a fundamental aspect of formal mathematics could play a revitalising role in mathematics education. To learn effectively, individuals need their own ethnomathematics valued and then, by acquiring academic mathematics, would be competent to choose for themselves the most appropriate to use in different circumstances. A wider multicultural perspective, where culture encompasses gender and class as well as ethnicity, can be used to help enable students to see how powerful mathematics can be as a tool for examining and operating within society. If this approach to multicultural mathematics is adopted, then the diverse aims of education may be more effectively fulfilled. A concomitant recognition within the curriculum of structural inequality in society would help to counterbalance present hegemonic forces and the resultant individualistic student-centred approaches to learning which have tended to normalise white male middle class activity and pathologise 'other'.

Many cultures have constructed their own ethnomathematics to meet their everyday needs. An emancipatory mathematics curriculum could validate each ethnomathematics whilst still acknowledging that many adults return to formal education to acquire the discursive practice and consequent rewards of academic mathematics. By denormalising academic mathematics and accepting the learners' various ethnomathematics, the curriculum can initiate processes of choice, control and empowerment.

Value systems of the curriculum

The roles of education in society include emancipation, distribution and socialisation. However, the goals of the curriculum frequently have a rhetoric located in emancipation but a reality located in distribution and socialisation. Whilst acknowledging that the curriculum must reflect the instrumental and life goals of the learners and the workforce requirements of society (skills and qualifications), there should be a more fundamental democratic basis for all learning. All curricula are constructed on a value system, but usually this is hidden, implicit and ideologically based. This value system needs to be made overt and explicit.

An obvious grounding for the value system lies in the notions of democracy and social justice. This would require education to be designed to meet individuals' learning requirements whatever their age, background, race, gender, disability or sexual orientation; draw on learners' educational values and experience; seek to develop their self-motivation and self-confidence; employ and promote co-operative rather than hierarchical approaches; and be about learning different ways of relating to and functioning within society. The value system would then be that of equity, social awareness, engagement and citizenship. It would require the empowerment of the learner as individual and citizen by developing mathematics to increase economic self-determination, overcome barriers to further and higher education and employment, and foster critical awareness and democratic citizenship.

These values could be used to generate problems and to encourage consideration of alternative mathematical techniques based on these values. This would enable learners to understand how knowledge is generated and established and critically relate this understanding to their own experiences.

However, rather than impose a value system on adult educators, an alternative and more powerful approach is for groups of mathematics educators to devise such 'mission statements' for themselves, as was done by the Massachusetts Adult Basic Education Team (Leonelli and Schwendeman 1994). A small group of tutors came together on a regular basis, wrote a mission statement, developed and made explicit a written belief system, and then devised *from* these a set of mathematics standards and implementation approaches. This innovative work in devising educational values and principles gives a model for the generation of a set of values and principles to underpin mathematics for mathematical education.

One of the real values of working with a group as described above is the opportunity to explore moral and ethical issues in a safe and supportive environment. What is clear is that if tutors do not construct a value system for themselves, the result will not be value-free mathematics education but education based on the value system of the current dominant political and educational forces.

Content of an emancipatory curriculum

The value and belief system discussed above will set the framework within which the rest of the curriculum – the content, pedagogy and contextual factors – will be shaped. Before going on to discuss these elements of an emancipatory curriculum, it is necessary to remember that the external forces identified throughout the book will be operating on the curriculum, institution, tutor and learner. It is likely that they will be pulling the curriculum in different directions. Nevertheless, it is important to establish alternative schema developed on alternative value systems as this contributes to the development of critical thinking amongst adult educators and exposes alternative paradigms. Without these, the current curriculum remains normalised and reproductive. The 'silent murmurings' remain unheard and mathematics remains for so many 'for other people'.

A constructivist model of knowledge and learning together with an awareness of Eurocentric bias in choosing the mathematics content and examples and activities which challenge the nature of formal mathematics, legitimates the ethnomathematics of 'other' cultures and exposes the myth of objectivity (Burton 1990). This could be the basis of a emancipatory mathematics curriculum where a critical understanding of numerical data prompts the questioning of unchallenged assumptions about how society is structured and enables the learner to act from a more informed position on societal structures and process (Frankenstein 1989; Abraham and Bibby 1992). As an emancipatory process, it could be very powerful in raising awareness that mathematics belongs to everyone; that the learner and teacher are engaged *together* in the learning and doing of mathematics and the world at large is accessible to analysis, criticism and transformation (Lerman 1992).

The education of adults requires instrumental learning, which allows control or manipulation of the environment through mathematics, *and* communicative learning, which involves values, intentions, feelings and moral decisions. Both are required in the curriculum. Instrumental learning, with its clearly defined needs and learning tasks, anticipated learning outcomes, behavioural objectives, competency-based education and measurable learning gains is well represented in the curriculum. What needs to be developed is the communicative learning with its emphasis on critical reflection of assumptions, discourse, validation of the learner's beliefs and reflective action upon the insights resulting from transformation of meanings (Mezirow 1994).

To open up mathematics in ways in which all can engage requires an active construction of understanding built on the learner's own knowledge and an exploration of the learner's own interests. The attempt to convey ideas and concepts to the learner must take place using the metaphors and imagery available to the learner. These are clearly the consequences of the society and culture within which the learner lives. This supports our earlier argument for the social construction of meaning in mathematics through the development of effective and participatory citizens committed to the support and development of their society. If mathematics is a dynamic, living and cultural product, the contextualisation of problems is essential.

Before reconceptualising the content of the adult mathematics curricula, we need to consider who designed the current curricula and why. No matter what the purpose of the learner, curricula are usually designed by academics in higher education for entry to the next academic level of mathematics, thus reflecting the pyramidal structure of formal education that is premised on failure and a process of exclusion. The arbitrary and mythical nature of this process is illustrated in the outcome of a conference on the mathematics requirements for Access courses at which admissions tutors in higher education were well represented (ACS 1991). It found that no list of mathematical topics has general acceptability for higher education entry and that present requirements for mathematics qualifications on entry to higher education are inconsistent with the nature of mathematics which admissions tutors identified as being of greatest importance.

Textbook writers, curriculum developers and, to a lesser extent, tutors are often white middle-class and male and this has led to a tendency to distort the curriculum which has systematically put certain groups at a disadvantage. Awareness of ethnicity, class and gender ensures rich contexts. This avoids alienation through the use of examples which use, say, white middle-class male-orientated activities to solve problems that have little to do with the learner's life and experiences. There are problems with contexts which require sensitivity on the part of the tutor. For example, is the use of contexts derived from women's traditional domains, say the home, reproductive or legitimising and hence emancipatory? Does it reclaim the domestic as a valid and valuable sphere of work or reinforce traditional expectations? These issues have to be brought out into the open and considered carefully.

A constant underpinning theme in the construction of both adult and mathematics education is relevance. This is often interpreted as locating problems in work or leisure pursuits. However, these quickly become outdated through changes in the workplace and society and provide such a wide variety of settings that choosing becomes difficult. Ironically this often results in just the irrelevance the curriculum is seeking to avoid. These points highlight the notion that the multiple numeracies used in different workplaces or different disciplines are all different *because* they are based in different discourses and practices and hence are specific to them. Hence mathematics needs to be learned and developed within the specific practice required and take into account the values and discourse of that practice. This suggests that to attempt to identify the content of the mathematics curriculum from the perspective of a suitable preparation for either work or further or higher education courses is likely to be ineffective.

In provision for adults, there is sometimes more freedom from prescribed curriculum. Here the ideas of genuine problem-generation around issues of critical citizenship and common interest such as the finances underpinning the privatisation of British Rail or a Housing Association budget can be constructed by groups of students and solved in a critical and analytical way. Issues such as pensions, for example, may be relevant to many adults in a way that is not true for school learning and involve judgements and mathematics well beyond percentages and compound interest. The tutor can share his/her mathematical knowledge but is just one of the group in terms of the ethical issues and value judgements. Here the body of mathematics knowledge is subsidiary. It becomes a library of accumulated experience. When a problem is generated which reveals a need for some of this knowledge, then the context, relevance and meaning of mathematics knowledge is established (Lerman 1992).

The recognition for the need to change the content of adult mathematics education is growing. Counihan (1994), reporting on a mathematics workshop provided for full and part-time adult students at the University of Southampton, found the sheer diversity of levels and abilities meant that there could be little collective work. The goals of the learners were often unclear but certainly not narrowly vocational. Adult mathematical education calls for more suitable materials for adults with graded goals and linked to some form of accreditation. It requires more than a vocational or recreational focus and requires inspirational mathematics to provide interest, intellectual challenge and confidence on which further study can be based. Similarly, the Final Report of an International Seminar on Adult Numeracy (CUFCO 1993) held in France concluded that there is a lack of teaching materials appropriate for adults learning mathematics and that this is having a serious effect.

An example of an inclusive humanising approach to mathematics is given by a course designed to teach women how to use calculus efficiently but also to help them see calculus as a human creation, developed in a particular cultural and historical context, by mathematicians who were influenced by the needs and values of the society in which they lived (Barnes and Coupland 1986). This course shows calculus not as some esoteric pure 'out-of-time' development but a response to the need of nations to extend their wealth and power through the accurate navigation which was crucial for exploration, colonisation and trade. The consequent importance of astronomy led Newton and Leibniz to invent calculus. The curriculum was constructed by negotiation with a pedagogy involving lectures to convey new topics and historical information, small group investigations to let learners explore new ideas and experiment with different methods, and both small group and whole class discussions to help learners clarify their thoughts, express concepts in their own words and so consolidate understanding. An important and interesting aspect of this course was the existence of some conflict between tutors and learners as to how each thought mathematics should be taught or learnt. This returns us to the thesis of this book that though these and other learners may see mathematics as culture and value-free, content, pedagogy and examples are not and cannot be politically neutral. Inequalities are enshrined and learners are a product of an unequal social system. Without an explicit awareness of this, we are in danger of being complicit in reproducing structural inequalities. Hence, an emancipatory curriculum needs not to support the *status quo* but to question and challenge it. The curriculum has the difficult task of both achieving this and meeting the institution's and the learner's need for control of the discourse of some part of formal mathematics and success in the qualification that recognises this.

A course which has been designed to teach empowering basic mathematics to adults in Spain uses a very standard syllabus of number, four basic operations, estimation/mental arithmetic, measurement, maps, proportions, data handling and statistics, and use of calculators but locates these in modules such as everyday economy, interpreting information, the world of work, consumer needs, health, technologies, environment, and social justice and democracy. For example, in interpreting information the four basic operations can be used to understand and question the distribution of seats in Parliament after Legislative Elections. In the world of work, the statistics used could be the probability of success in a job application based on the number of candidates and the number of vacancies. In environment, estimation could concern the amount of water saved at home when restrictions are used. In social justice and democracy, calculators may be used to understand the specifications of an opinion poll and its reliability (Plaza 1996).

Learning and teaching of emancipatory mathematics

When mathematics is seen as a social construct interwoven harmoniously with cultural contexts, then the basis of the pedagogy must lie in the belief that learners develop mathematics concepts within this context through problem-posing activities generated through the learners' own concepts, values and processes. Mathematics and mathematics learning develops from important activities within the cultural context and both are social constructs based on values and beliefs arising from problem-solving and investigation. As the epistemological base is from the learner's own culture and contexts, this provides motivation and confidence and hence is emancipatory. The pedagogy can be developed to assist the learner construct knowledge using their own cultural context as starting point. The advantages are clearly motivation, confidence, self-esteem, increased control and hence empowerment.

This pedagogy does, however, make great demands on the tutor in terms of secure mathematical and personal awareness to allow this exploration and challenge of ideas. It also requires the tutor to have an understanding and knowledge of mathematics as an activity that occurs in social and cultural contexts (Neyland 1995). Difficult though this pedagogy is, the alternative has been singularly ineffective. Teaching based on passive didactic modes of learning are unlikely to get learners to examine or change their own prejudices, assumptions and stereotypes. Learning is most effective when based on active participation through discussion, based on experiences and reflection. This allows the learner to draw on their own experiences as real knowledge. Differences in experiences and backgrounds allows different voices to be heard in the learning conversation and to expose the different perspectives on the world. Hence, the pedagogy should be based on a variety of learning approaches but involve co-operative projects, problem-solving and discussion applied to the thorough and rigorous treatment of mathematics required for external assessment. The teacher's role is the difficult one of respect for the learner whilst acknowledging the power asymmetries of all formal learning.

Materials need to be screened for bias and stereotyping and where these occur, they can be discussed and made part of the educational exercise. Learners become aware that struggle is not always to be avoided and confidence is increased by overcoming obstacles, not avoiding them. This pedagogy would make serious demands on the tutor who may have little mathematics training themselves, be from other disciplines, have little or no formal training in adult education or even be teaching mathematics unwillingly. Realistically, implementation of this pedagogy calls for systematic and rigorous education for the adult education mathematics tutor.

There is limited training and support for tutors who teach adults mathematics. There is such a wide range of experience and expertise of those either engaged or wishing to engage, in this work that a range of programmes is required to meet their differing needs. Such a range of distance learning materials designed to train and support teachers of adult numeracy and mathematics has been developed in Australia. The packages include *Breaking the Maths Barrier* by Marr and Helme (1991) and *Adult Numeracy Teaching*, a National Staff Development Committee initiative (1995). The latter is an 84-hour pack which has the aim of blending theory and practice about teaching and learning adult numeracy within a context of doing and investigating some mathematics, whilst developing a critical appreciation of the place of mathematics in society. It is rigourous, entertaining, grounded in both up-to-date practice and theory. It would enable any tutor who had completed the course to facilitate empowering mathematics in their own students. But, just as importantly, in an area where many tutors themselves may lack confidence in their mathematical and/or tutoring abilities, it would empower the tutors themselves.

Because of its importance to the pedagogy of an emancipatory curriculum, part of the earlier discussion on the notion of learning as the acquisition of the discourse of the subject will be reiterated. The underlying motivation driving adult students to learn mathematics in a formal situation is to gain access to the powerful and prestigious discourse of academic mathematics. This approach helps to explain why they are less interested in building on their own existing everyday mathematical discourses which they feel are not valued in society. Within our society, esteem or value is only given to those who grasp the correct forms of knowledge and ways of constructing that knowledge. These are rewarded with qualifications. Those who gain 'other' knowledge (ethnomathematics) are not given qualifications. Though able to solve problems, if this is not expressed in the discourse of formal mathematics, it is not valued.

One of the most significant features of the process of knowledge being owned, both in its practice and its definition, by the academy, is the way that adults who have a wealth of experience see their own knowledge as somehow being different and other to 'real' knowledge which only institutional teaching gives them (Stuart 1995). So adults come to formal learning wishing to move themselves away from their local mathematics and, by gaining entry to the discourse of formal mathematics, operate in the world in new ways and gain social power and financial advantage through being able to speak as an insider to this discourse. Learners need pathways from their everyday discourses into the unfamiliar terrain of the academic discourse. Teaching and the syllabus as a whole can be thought of as narratives which develop from the familiar, perhaps through case studies, and are plotted to lead students into the unfamiliar so that they may become users of the new discourse themselves. The use of metaphor allows the learner to position themselves with regard to the new knowledge.

If adults learn mathematics in part to acquire this communicative competence in mathematics language, then techniques are required to direct the learner's attention to the nature of the discourse while still retaining a normal level of communication. This requires the use of both commenting and meta-commenting by the tutor whereby the comments deal with the mathematics under discussion and the metacomments draw attention to issues of the discourse itself. Knowledge of this discourse can be systematically developed by dialogue not only between teacher and student but also between students through group work. By working with others while learning mathematics, students can, through discussion and dialogue, break patterns of dependency, social relationships and isolated learning.

The pedagogy will be required to develop other skills alongside competence in mathematical techniques. These will include critical thinking, control over the discourse of mathematics and communication skills. Critical thinking involves the two inter-related processes of identifying and challenging assumptions and imagining and exploring alternatives. This involves a continuous process composed of alternate phases of reflecting on a problem, testing new solutions, strategies or methods on the basis of that thinking, reflecting of the success of these actions in particular contexts and further honing, refining and adapting these actions according to alternative contexts.

Action may be mental or behavioural. The learner starts with contextually-grounded ideas about what works well in particular contexts, particularly their own context and ethnomathematics. They develop an explanation of why the ethnomathematics works well in its context and then a readiness to alter these practices according to changing contexts. This allows the learner to move from their ethnomathematics to academic mathematics. Learner satisfaction is not the sole aim of critical thinking and there may be frustration and struggle. Critical thinking is often a difficult and sometimes painful process (Brookfield 1987).

If the learner is to use mathematics effectively in everyday situations, they will need to be able to express mathematical ideas in a communicable written form. If they are to design their own mathematics problems, they need to be able to write mathematics. So the language of mathematics requires reading, talking and writing as well as mathematical skills but these need to be developed through practice. Problem generation, by individuals or preferably small groups, from areas of current interest to the learners can be first formulated verbally then written out. If the problem is from real life, it is likely that the context-setting may be long. This encourages the learner to see mathematics problems as an integral, but sometimes small, part of a wider problem. Other students can then work through this problem, hence giving practice in reading and discussing mathematics.

Keeping a personal journal of learning on a mathematics course can encourage the learner not just to examine the mathematics learnt but also to critically reflect on their experiences of the course. By reflecting on their learning, the student understands more clearly their mathematics anxiety, mathematical concepts and techniques and which learning styles suit them best. The journal can also help the learner to discover the solution to problems through the process of thinking about them in writing. The consequent organisation of ideas and notions about the topic or problem, though sometimes difficult, makes connections across the subject.

The emancipatory pedagogy advocated here sees the process of learning as active and problem-centred, not passive or learned from books. The knowledge valued is the personal local or ethno-knowledge gained from real experience rather than absolute expert or academic knowledge gained from textbooks (though the eventual control of the latter knowledge is seen as an objective). Learning is seen as through action or investigation thus constructing new knowledge rather than learning which emphasises that the knowledge exists independently of the learner. The relationship with the tutor is based on egalitarianism rather than authoritarianism and control. Assessment is based on self-, peer- and tutor-collaboration rather than unilateral tutor judgement (McConnell 1995).

The pedagogy starts where it seems appropriate, using discursive methods, investigations, reflection through diaries, group work, self-paced study in a warm, friendly but demanding environment, appropriate assessment and group projects; all founded on a critical assessment of whatever is being studied. Both problem-solving and investigation can be used, though a clear distinction needs to be made between them. In the former, the tutor poses a problem, leaving the learner to find their own way to solve it. In the latter, the tutor chooses the starting point or agrees the learner's choice, then the learner defines the problem within the given situation and attempts to solve in own way. The first aims to find a solution; the second to open up the problem to encourage and cultivate the learner's curiosity rather than trying to lead them to reach a particular conclusion or go down a particular path. The emancipatory pedagogy presents a range of alternative experiences and opinions backed by insight into how to arrive at conclusions critically and to make informed choices. It respects and legitimates learner autonomy, personal knowledge, enquiry-based learning and collaborative organisation.

Within this pedagogy, the tutor's attitude is crucial. It needs to be premised on encouragement, support, enthusiasm, high expectation and a fundamental assumption that the mathematics classroom should be a place where originality and independent and creative thinking are valued and common sense knowledge or ethnomathematics validated and built on, not rejected as irrelevant and unimportant.

Support for at least some of the notions underlying this pedagogy comes from a recent investigation of mathematical competence in primary schools in Britain, Germany and Switzerland which found that the Continental countries gave children a sounder and more efficient mathematics education. Various factors emerge to explain this but the report particularly noted the greater emphasis on mental arithmetic, oral work group discussion and student constructed methods. More time is allocated to securing the ground before moving on to the next concept. There is a widely held view on the continent that mental calculation promotes pupils' conceptualisation of number and that mathematics provides a unique subject for training the minds of pupils in extended chains of reasoning (Bierhoff 1996)

Environment for an emancipatory curriculum

A fundamental requirement for an emancipatory curriculum is an operational and effective institutional equal opportunities policy. A statement is not sufficient. The factors in institutional policy likely to lead to success in enhancing equality of opportunity have been identified as clear and public formulations of long-term aims with plans for their achievement, the setting of short-term goals with timescales and a wide ownership of policy (Farish *et al* 1995). Structures found to facilitate equal opportunities initiatives include active staff and student participation in developments, appropriate funding and resources, user-friendly procedures, monitoring, evaluation and, most importantly, a pervasive equal opportunities ethos.

The equating of educational disadvantage with deficit or deprivation leads to the concept of compensatory education. Viewing disadvantage as a result of structural inequalities, on the other hand, results in a perceptive transformation which is the requirement for emancipatory empowering education. A move to a more constructivist approach to mathematics allows the knowledge and social experience of the learner to be reflected back as valid and significant and the mathematical knowledge hidden in the family, work or community to be recognised. From this the learner can then move to a wider awareness. The underlying assumptions are that low motivation and achievement are closely related to alienation, that adults are capable of undertaking sustained demanding education if it is seen as relevant to their needs, and that the mutual dialogue which incorporates cultural interests changes the opportunity to learn.

Diversity of experience enriches our cultural understanding and minority voices need to be heard. Hence there is a need for ways of changing intentional or unintentional discriminatory institutional practices and procedures. There is a need to incorporate exemplars from cultures and traditions that have been ignored or devalued for too long. The multicultural approach can be seen as part of the general strategy of making mathematics more accessible to all. A Cultural Policy can be developed on an institutional basis built on the experiences of Gender Awareness Policies. This might include an institutional commitment to monitor:

- the economic and social context in the area in which the provision will be offered,
- the social construction of culture within the cultural, religious and communal traditions of the community,
- both practical needs (*eg*, child care) and strategic needs (*eg*, ways of challenging traditional roles),
- a critical diagnosis of the institutional/educational framework,

- if necessary, the development of educational programmes with culture specific aims and objectives.

This critical reflection on culture can be an integral part of the institution's strategic plan and, within this, the adult educator's mathematics curriculum. The latter could incorporate a discussion of the socially constructed nature of mathematics knowledge and hence its fallibility, the multicultural origins of mathematics and the validity of the mathematics of all cultures.

The unequal nature of society and institutions could be illustrated and condemned, the social context of all education discussed and the relationships between power and knowledge exposed. This would lay the groundwork for an analysis of resultant problems encountered by certain groups in society. This could lead to an exploration of the position of mathematics and statistics in our society and the role of mathematics in controversial and value-based issues. The notion of academic mathematics, multicultural mathematics and ethnomathematics could be made explicit through discussion. The elimination of overt and covert prejudice from text books, other curriculum materials and examinations and the widening of the range of applications to everyday or socially valuable examples, eg appliances for the disabled, would help.

But equal opportunities policies have an additional fundamental role to play in a democratic society. Citizenship has to be learnt like any other skill but it will be learned not through the formal curriculum but through positive experiences of participation. Participatory democracy is learned through practice and therefore the adult education experience should itself be an experience of participatory democracy. In this way it can be an affective as well as cognitive learning experience that both democracy and adult education are 'for us' and not just 'for other people'.

Practical considerations that are required for the environment of an emancipatory curriculum have been well documented in reports from NIACE (see, for example, NIACE Replan 1991). These include awareness of learners' lifestyles, constraints, family responsibilities and commitments; provision of good quality childcare and accessible, welcoming and safe premises; and careful consideration of financial matters such as controlled enrolment fee levels, remission schemes and information and advice on implication of study for benefits. Networking and outreach work is necessary to ensure that provision is geared to local need. Publicity should be clear, jargon-free and easy to understand. Free, accessible and impartial educational guidance is crucial.

Conclusion

This book has considered the varied forces acting upon adults attempting to learn mathematics, tutors teaching adults mathematics and adult mathematics curricula. It has examined how the dominant political ideology, the current ideas about the nature of mathematics, educational philosophies and the structure of the society in which we live, all operate on the learner, tutor and curriculum to locate learning mathematics on a continuum from an empowering emancipatory experience to a destructive deadening one. This final chapter has outlined some notions that might contribute to an emancipatory curriculum for adults but has not discussed the detail of the content because this will vary.

The book has argued that an emancipatory curriculum must be based on notions of respect for all learners, social justice, democracy and the passionate belief that to be in control of mathematics is a step towards being in control of one's life. These beliefs should inform and direct the curriculum whatever mathematics topic is under consideration. However, adults' expressed needs should not be ignored and any widening of the curriculum should not replace the instrumental goals and self-development requirements of the learner but enhance these. Adults can be encouraged to recognise and value the mathematics learning that takes place in all facets of their everyday life from individual to worker to citizen.

The result of such a curriculum is that the learner will acquire the skills and qualifications in mathematics that they need for social and economic advancement but also the critical awareness required to change or attempt to change restrictions acting upon them. The informal ethnomathematics of the various groups in society exist within society alongside the professional skills of privileged groups. Which of these is valued is a political decision which involves all of society. The choice of which knowledge should be canonised in the curriculum and how this curricular knowledge relates back to society is a political decision which in a democratic society should seek to provide for the needs of all citizens.

I should like to conclude with a quote from an adult education tutor who sums up the potential outcome of an emancipatory curriculum:

> by making progress with a subject they had thought they were unable to do, they gained an enormous sense of achievement and increased confidence which informed their whole approach to study and life.

Bibliography

This bibliography contains a minority of items which, while not referenced in the text, have provided pertinent background to the discussion.

Abraham, J. and Bibby, N. (1992) 'Mathematics and society: ethnomathematics and the public educator curriculum' in M.Nickson and S.Lerman (eds) pp.180–201.

ACS (1991) *Access: Implications for Mathematics*, London: ACS.

Adams, C. and Walkerdine, V. (1986) *Investigating Gender in the Primary School: Activity-based Inset Materials for Primary Teachers*, London: Inner London Education Authority.

Advisory Council for Adult and Continuing Education (1982a) *Adults: Their Educational Experiences and Needs*, Leicester: ACACE.

Advisory Council for Adult and Continuing Education (1982b) *Continuing Education: From Policy to Practice*, Leicester: ACACE.

Advisory Council for Adult and Continuing Education (1982) *Adults' Mathematical Ability and Performance*, Leicester: ACACE.

Ainley, J. (1991) 'Playing games and real mathematics' in Pimm, D. (1991) pp.239–248.

ALBSU (1992) *A Survey of Literacy and Numeracy Students*, London: ALBSU.

ALBSU (1993) *Basic Skills Support in Colleges*, London: ALBSU.

ALBSU (1994) *Basic Skills in Everyday Life*, London: ALBSU.

Anderson, R. and Darkenwald, G. (1979) *Participation and persistence in American adult education*, Papers in Lifelong Learning, College Entrance Examination Board.

Anderson, S. E. (1990) 'Worldmath curriculum: fighting Eurocentrism in mathematics' *Journal of Negro Education*, 59(3), pp.348–359.

Apple, M. (1982) *Education and Power*, Boston: Routledge and Kegan Paul.

Argyris, C. and Schon, D. (1974) *Theory in Practice: Increasing Personal Effectiveness*, San Francisco: Jossey-Bass.

Ascher, M. (1991) *Ethnomathematics: a Multicultural View of Mathematical Ideas*, California: Brooks/Cole Publishing Co.

Ashley, D. and Betebenner, D. (1993) 'Mathematics, postmodernism, and the loss of certainty' Paper presented to the British Sociological Association Annual Conference, University of Essex.

Asimov, I. (1977) *Asimov on Numbers*, New York: Pocket Books.

Avari, B. (1995) 'Multicultural education for adults' in Bryant, I. (1995) pp.12–17.

Bagnall, R. (1991) 'Relativism, objectivity, liberal adult education and multiculturalism' *Studies in the Education of Adults*, 23(1), pp.61–84.

Barnes, M. and Coupland, M (1986) 'Humanising calculus: a case study in curriculum development' in Burton, L. (1986).

Barnes, M., Johnston, B. and Yasukawa, K. (1995) 'Critical Numeracy', a poster presented at the Regional ICME Conference, Monash University.

Barnett, R.(1990) *The Idea of Higher Education*, Buckingham: SRHE and Open University Press.

Barr, G. (1993) 'Numeracy: a core skill or not?' *Viewpoints 16: Numeracy*, London: ALBSU.

BBC Broadcasting Research (1990) *Basic Skills: Numeracy*, London: BBC.

Belenky, M., Clinchy, B., Goldberg, N. and Tarule, J. (1986) *Women's Ways of Knowing*, New York: Basic Books.

Ben Tovin, G. *et al* (1986) *The Local Politics of Race*, London: Macmillan.

Benn, R. (1994) 'Access tutors do it without training: an investigation of professional development needs in adult education', in Benn, R. and Fieldhouse, R. (eds), *Training and Professional Development in Adult and Continuing Education*, Centre for Research in Continuing Education, University of Exeter Occasional Paper 1, pp.11–17.

Benn, R. and Burton, R. (1993) 'Access mathematics: a bridge over troubled waters', *Journal of Access Studies*, 8(2), pp.180–190.

Benn, R. and Burton, R. (1994a) 'Participation and the Mathematics Deterrent', *Studies in the Education of Adults*, 26(2), pp.236–249..

Benn, R. and Burton, R. (1994b) 'Women's experience of mathematics and access to higher education in Britain: just another brick in the wall?', in M. Woodrow (ed) *Access to Higher Education: The Economic and Social Implications: An East/West Convention*, Department of Education, The Netherlands.

Benn, R. and Burton, R. (1995a) 'Access and targeting: an exploration of a contradiction', *International Journal of Lifelong Education*, 14(6), pp.444–458.

Benn, R. and Burton, R. (1995b) 'Mathematics: a peek into the mind of God' in Almeida, D. and Ernest, P. (eds) *Perspectives on Mathematics*, Exeter: University of Exeter Perspectives Series, pp.7–17.

Benn, R. and Burton, R. (forthcoming), An Investigation of Ethnic Minority Participation On Access Courses.

Benn, R. and Fieldhouse, R. (1990) 'Adult education to the rescue in Thatcherite Britain' in Poggler, F. and Yaron, K. *Adult Education in Crisis Situations*, Jerusalem: The Magnes Press pp.73–86.

Benn, R. and Fieldhouse, R. (1993) 'Government policies on university expansion and wider access, 1945–1951 and 1985–1991 compared' *Journal of Studies in Higher Education* , 18(3) pp.219–313. 3

Benn, R. and Fieldhouse, R. (1994) 'Raybouldism, Russell and New Reality' in Armstrong, P., Bright, B. & Zukas, M. (eds). *Reflecting on Changing Practice, Contexts and Identities*, Hull: SCUTREA, pp. 7–9.

Benn, R. and Fieldhouse, R. (1995) 'Adult education and National Independence Movements in West Africa, 1947–1953, and some lessons for the present day' in Poggeler, F. (ed) *National Identity and Adult Education*, Frankfurt am Main: Peter Lang, pp.247–260.

Bernstein, B. (1970) 'A critique of the concept of compensatory education' in Rubenstein, D. and Stoneman, C. (eds) *Education for Democracy*, Harmondsworth: Penguin.

Beveridge, L. (1995) 'Using reflective journals in numeracy classes' in Coben, D. (comp.) (1995) pp.117–118.

Bierhoff, H. (1996) *Laying the Foundations of Numeracy*, London: NIESR

Bishop, A. (1985) 'The Social psychology of mathematics education' Plenary paper presented at the Ninth P.M.E. Conference, Noordwijkerhout, Holland, July, 1985 quoted in Presmeg, N. (1988).

Bishop, A. (1990) 'Western mathematics: the secret weapon of cultural imperialism' *Race and Class*, 32(2), pp.51–65.

Bishop, A. (1991) 'Mathematics education in its cultural context' in Harris, M. (1991) pp.29–41.

Bjoerkqvist, O. (1996) 'Critical Education' a paper given to the 8th International Conference on Mathematics Education, Seville, Spain.

Black, S. (1995) 'Knitting tensions: the prescriptive versus the visual' in Coben, D. (1995b), pp. 49–52.

Bloor D. (1973) 'Wittgenstein and Mannheim on the sociology of mathematics' *Studies in the History and Philosophy of Science*, 4(2), pp.173 – 191.

Bloor, D. (1976) *Knowledge and Social Imagery*, London: Routledge and Kegan Paul Ltd.

Bourdieu, P. (1971) 'Systems of education and systems of thought' in Young, M. (ed) *Knowledge and Control: New Directions in the Sociology of Education* London: Collier-Macmillan quoted in Westwood, S. (1981).

Breen, C. (1990) 'Reflecting on energy activities: an attempt at liberating pre-service mathematics teachers at a South African university' in R. Noss *et al* (1990) pp.36–45.

Bright, B. (ed) (1989) *Theory and Practice in the Study of Adult Education: The Epistemological Debate*, London and New York: Routledge.

Bright, B. (1992) 'Theories-in-use, reflective practice and the teaching of adults: professional culture in practice' in Miller, N and West, L. (1992) pp. 24–27.

Brookfield, S. (1980) *Independent Adult Learning*, Unpublished Doctoral Thesis, Leicester: University of Leicester.

Brookfield, S. (1987) *Developing Critical Thinkers*, Milton Keynes: Open University Press.

Brookfield, S. (1989) 'The epistemology of adult education in the United States and Great Britain: A cross-cultural analysis', in Bright, B. (1989) pp.141–173.

Broverman, I. *et al* (1970) 'Sex-role stereotypes and clinical judgements of mental health' *Journal of Consulting and Clinical Psychology*, 34(1), pp.1–7.

Brown, R. (1989) 'Carpentry: a fable' *Mathematical Intelligencer*, 11(4), p.37.

Bryant, I. (ed) (1995) *Vision, Invention, Intervention*, University of Southampton: SCUTREA.

Buerk, D. (1981) 'The voices of women making meaning in mathematics' *Journal of Education*, 167(3), pp.59–70.

Burton, L. (1986) *Girls Into Maths Can Go*, London: Holt, Rinehart and Winston.

Burton, L. (1989) 'Images of Mathematics' in P. Ernest (1989) pp.180–187.

Burton, L. (1990) 'Passing through the mathematics critical filter-implications for students, courses and institutions' *Journal of Access Studies*, 5(1), pp.5–17.

Burton, L. (ed) (1990) *Gender and Mathematics*, London: Cassell Education.

Buxton, L. (1984) *Do You Panic About Maths?*, London: Hienemann.

Candy, P.C. (1981) *Mirrors of the Mind: Personal Construct Theory in the Training of Adult Educators*, Manchester: Manchester Monographs.

Carraher, D. (1991) 'Mathematics in and out of school: a selective review of studies from Brazil' in Harris, M. (1991) pp.169–201.

Carraher, T and Schliemann, A. (1988) 'Culture, arithmetic, and mathematical models' *Cultural Dynamics*, 1(2) pp.180–194 quoted in Carraher, D. (1991).

Clarke, P. (1993) 'Putting class back on the agenda' *Journal of Access Studies*, 8(2) pp. 225–230.

Coaker, P.B. (1985) 'Why teach mathematics?' *1985 Presidential Address: The Mathematical Gazette*, 69(449), pp.161–174.

Coats, M, (1994) *Women's Education*, Buckingham: SHRE and Open University Press.

Cobb, P. (1986) 'Contexts, goals, beliefs and learning mathematics' *For the Learning of Mathematics*, 6(2), pp.2–9.

Coben, D. (comp) (1995a) *Adults Learning Mathematics-A Research Forum: Proceedings Of The Inaugural Conference 1994*, London: Goldsmiths College.

Coben, D. (comp) (1995b) *Mathematics with a Human Face*, London: Goldsmiths, University of London.

Coben, D. and Thumpston, G. (1995) 'Mathematics life histories: guilty secrets and untold joys' in Hoar, M. *et al* (1995) pp.38–41.

Cockcroft, W. H. (Chairman of the Committee of Inquiry into the Teaching of Mathematics in Schools) (1982) *Mathematics Counts*, London: HMSO.

Collins, M. (1991) *Adult Education as Vocation: A Critical Role for the Adult Educator*, London and New York: Routledge.

Collins, M. (ed.) (1995) *The Proceedings of the International Conference on Educating the Adult Educator: The Role of the University*, College of Education, University of Saskatchewan

Colwell, D. (1995) 'A discussion exploring areas of enquiry into the role of language in learning mathematics' in Coben, D. (1995a)

Commission for Racial Equality (1990) *Code of Practice for the Elimination of Racial Discrimination in Education*, London: CRE.

Commission on Citizenship (1990) *Encouraging Citizenship*, London: HMSO.

Cornelius, M. (1992) *The Numeracy Needs of New Graduates in Employment*, University of Durham.

Corrigan, P. (1992) 'The politics of Access courses in the 1990s' *Journal of Access Studies*, 7(1), pp.19–32.

Counihan, M. (1994) 'Making mathematics accessible' *Adults Learning*, 5(7), pp.179–181.

Cross, K. (1990) 'Sharing perspectives: people learning mathematics' *Mathematics Teaching*, 130.

Crowther Report (1959) *15–18: Report of the Central Advisory Council of Education (England) vol 1*, London: HMSO.

CUFCO *The Final Report of an International Seminar on Adult Numeracy* France: European Network for Research, Action and Training in Adult Literacy and Basic Education.

Daines, J., Daines, C. and Graham, B. (1992) *Adult Learning Adult Teaching*, Department of Adult Education, University of Nottingham.

D'Ambrosio, U. (1990) 'The role of mathematics in building up a democratic society and the civilizatory mission of the European powers since the discoveries' in R.Noss *et al* (1990) pp.13–21

D'Ambrosio, U. (1991) 'Ethnomathematics and its place in the history and pedagogy of mathematics' in Harris, M. (1991) pp.15–25.

Davis, P.J. (1986) 'Fidelity in mathematical discourse: is one and one really two?', in Tymoczko, T. (ed) *New Directions in the Philosophy of Mathematics*, Boston: Birkhauser.

De Corte, E. and Verschaffel, L. (1991) 'Some factors influencing the solution of addition and subtraction word problems' in Durkin, K. and Shire, B. (ed) *Language in Mathematical Education, Research and Practice*, Buckingham: Open University Press, pp.117–130.

Denny, J.P. (1986) 'Cultural ecology of mathematics: Ojibway and Inuit hunters' in Closs (ed) *Native American Mathematics*, Austin: University of Texas Press.

Department of Education and Science (1973) *Adult Education: A Plan for Development* London: HMSO.

Department of Education and Science (1987) *Higher Education: Meeting The Challenge*, London: HMSO.

Department of Education and Science (1988) *Final Report of the Mathematics Working Group*, London: HMSO.

Department of Education and Science (1991) *Education and Training for the 21st Century*, London: HMSO.

Department of Education and Science (1991) *Higher Education: A New Framework*, London: HMSO.

Desforges, C. (1989) 'Classroom processes and mathematical discussions: a cautionary note' in Ernest, P. (1989) pp.143–150.

Dewdney, A. (1993) *200% of Nothing*, New York: John Wiley and Sons Inc.

Dewey, J. (1964) *Democracy and Education*, London: Macmillan.

DfEE, the Scottish Office and the Welsh Office (1995) *Lifetime Learning – A Consultation Document*, Sheffield: DfEE.

Dodd, A. (1992) 'Insights from a maths phobic' *The mathematics teacher* 85(4) April pp.296–298

Dorn, A. (1991) 'Ethnic minority participation in higher education' Unpublished paper to PCFC quoted in Bird, J. *et al* (1992) 'Rhetorics of access – realities of exclusion?' *Journal of Access Studies*, 7(2), pp.146–163.

Dowling, P. (1991) 'The contextualisation of mathematics: towards a theoretical map' in Harris, M. (1991) pp.93–120.

Duke, C. (1992) 'Adult education, social movements and democracy: what connections?' in

Proceedings of the International Conference on Adult Education as a Social Movement, Wroclaw.

Duke, C. (1992) 'Learning Citizenship' *Adults Learning*, 4(2), pp.49–52.

Eastway, R. (1995) 'What's the difference between a puzzle and a maths question?' in Coben, D. (1995b) pp.152–155.

Edwards, R and Usher, R. (1995) 'Postmodernity and the education of educators' in Collins, M. (ed.) (1995) pp.61–74.

Edwards, R. (1995) 'Behind the banner: whither the learning society?' *Adults Learning*, 6(6) pp.187–189.

Edwards, R. and Usher, R. (1994) 'Tribes and Tribulations: narratives and the multiple identities of adult educators' in Armstrong, P. *et al* (eds) *Reflecting on Changing Practices, Contexts and Identities* University of Hull: SCUTREA pp.32–35.

Emblen, V. (1991) 'Asian children in school', in Pimm, D. (1991).

Ernest, P. (1986) 'Games: a rationale for their use in the teaching of mathematics in school' *Mathematics in School*, 15(1) pp.2–5.

Ernest, P. (1990) 'Aims in mathematics education as an expression of ideology' in R. Noss *et al* (1990) pp.85–92

Ernest, P. (1991) 'Mathematics, the public educator perspective and critical citizenship' in Thorstad, I. (ed) *Proceedings of a Seminar on Adult Numeracy*, Essex: Mathematics Department, University of Essex, pp.17–20.

Ernest, P. (1991) *The Philosophy of Mathematics Education*, Basingstoke: Falmer

Ernest, P. (1992) 'The National Curriculum in mathematics: its aims and philosophy' in Nickson, M. and Lerman, S. (1992) pp.33–61.

Ernest, P. (1995) 'The nature of maths and teaching' in Almeida, D. and Ernest, P.(eds) *Perspectives on Maths*, Exeter: University of Exeter Perspectives 53, pp.29–41.

Ernest, P. (ed) (1989) *Mathematics Teaching: the State of the Art*, Falmer: Lewis.

Evans, J. (1990) 'Mathematics learning and the discourse of critical citizenship' in R. Noss *et al* (1990) pp.93–95.

Evans, J. (1991) 'Numeracy, mathematics and critical citizenship' in Thorstad, I (comp) *Proceedings of a Seminar on Adult Numeracy*, Research Reports, Department of Mathematics, University of Essex, pp.21–26.

Evans, J. (1992) 'Mathematics for adults: community research and "barefoot statisticians"' in M. Nickson and S. Lerman (1992) pp.202–216.

Evans, J. and Harris, M. (1991) 'Theories of Practice' in Harris, M. (1991) pp.202–210.

Farish, M. *et al*, (1995) *Equal Opportunities in Colleges and Universities: Towards Better Practices*, Buckingham: SRHE and Open University Press.

Fasheh, M. (1991) 'Mathematics in a social context: math within education as praxis versus math within education as hegemony' in Harris, M. (1991) pp.57–61.

Fauvel, J. (1990) 'Mathematics and empowerment: historical perspectives' in R. Noss *et al* (1990) pp.104–105.

Fennema, E. (1979) 'Women and girls in mathematics – equity in mathematics education' *Educational Studies in Mathematics*, 10(4), pp.389–401.

Fennema, E., Peterson, P., Carpenter, T. and Lubinski, C. (1990) 'Teachers' attributions and beliefs about girls, boys and mathematics' *Educational Studies in Mathematics*, 21 pp.55–69.

Field, J. (1992) 'Utopia and Adult Education' *Oswiata Doroslych Jako Ruch Spoleczny* Wroclaw, pp.141–156.

Field, J. and Weller, P. (1993). 'Vocationalising Access? Training and Enterprise Councils and the Access movement', *Journal of Access Studies*, 8(1), pp.27–38.

Fieldhouse, R. (1985) 'The problems of objectivity, social purpose and ideological commitment in university adult education' in Taylor, R., Rockhill, K. and Fieldhouse, R. (1985) pp.29–51.

Fieldhouse, R. (1993a), *Optimism and Joyful Irreverence: The Sixties Culture and Its Influence on British University Adult Education and the* WEA, Leicester: NIACE.

Fieldhouse, R. (1993b) 'Have we been here before? a brief history of social purpose adult education' *Adults Learning*, 4(9), pp.242–243.

Fieldhouse R. and Taylor R. (1988) 'Ideology and political socialisation in the United Kingdom' in Fieldhouse R. (ed) , pp.1–18.

Fieldhouse R. (ed) (1988) *The Political Education of the Servants of the State*, Manchester: MUP.

Flax, J. (1990) *Thinking Fragments: Psychoanalysis, Feminism and Postmodernism in the Contemporary West*, University of California Press.

Foden, F (1992) *The Education of Part-Time Teachers in Further and Adult Education: A Retrospective Study*, Leeds: Leeds Studies in Continuing Education.

Foucault, M. (1973) *The Order of Things: An Archaeology of Human Sciences*, New York: Vintage Books.

Foucault, M. (1977) *The Archaeology of Human Knowledge*, New York: Vintage Books.

Foucault, M. (1986) *Power/Knowledge*, Brighton: The Harvester Press.

Frankenstein, M. (1989) *Relearning Mathematics: A Different Third R – Radical Maths*, London: Free Association Books.

Frankenstein, M. (1990) 'Critical mathematics literacy' in R.Noss *et al* (1990) pp.106–114.

Freire, P. (1972) *Pedagogy of the Oppressed*, Harmondsworth: Penguin

Freire, P. (1985) *Cultural Action for Freedom*, Harmondsworth: Penguin.

Frye, S. (1982) Contribution to the Proceedings of the 4th International Congress on Mathematical Education quoted in Corneilius, M.(1991) 'What do we know about graduates and adult numeracy' in Thorstad, I (1991) (comp) *Proceedings of a Seminar on Adult Numeracy*, Research Reports, Department of Mathematics, University of Essex.

Further Education Unit (1987) *Marketing Adult Continuing Education: A Project Report*, FEU.

Gal, I. (1996) 'Statistical literacy: the promise and the challenge' a paper given to The 8th International Conference on Mathematical Education, Seville, Spain.

Gerdes, P. (1986) 'On culture: mathematics and curriculum development in Mozembique' in Johnsen-Hoines, M and Mellin-Olsen, S. (eds) *Mathematics and Culture*, Bergen: Caspar Forlag, Bergen Laergerhogskole.

Gerdes, P. (1996) 'Culture and mathematics education in (southern) Africa' a paper given to The 8th International Conference on Mathematical Education, Seville, Spain.

Gilligan, C. (1982) *In a Different Voice*, Cambridge: Harvard University Press.

Ginsburg, H. and Russell, R. (1981) *Social Class and Racial Influences on Early Mathematical Thinking*, Chicago: SRCD, p.56.

Ginsburg, L. and Gal, I. (1996) 'Uncovering the knowledge that adult learners bring to class' a paper given at The 8th International Conference on Mathematical Education, Seville, Spain, July 1996.

Girls and Mathematics Unit (1988) *Girls and mathematics: Some Lessons for the Classroom*, London: Economic and Social Research Council.

Giroux, H.A. (1985) 'Critical pedagogy, cultural politics and the discourse of experience' *Journal of Education*, 167(2).

Government Statisticians Collective (1979) 'How official statistics are produced: views from the inside' in Irvine, Miles and Evans (eds) *Demystifying Social Statistics*, London: Pluto Press.

Gramsci, A. (1971) *Selections from the Prison Notebooks*, London: Lawrence and Wishart.

Griffin, C. (1983) *Curriculum Theory in Adult and Lifelong Education*, London: Croom Helm.

Grinter, R. (1985) 'Bridging the gulf', *Multicultural Teaching*, 3(2).

Groombridge, B. (1983) 'Adult education and the education of adults' in Tight, M. (1983) *Adult Learning and Education*, London: Routledge, pp.3–19.

Halsey, A.H. (1975) Foreword in Lovett, T. (1975)

Halsey, A.H. (1992) An international comparison of Access to higher education in Phillips, D. (ed) *Lessons of cross-national comparison in education*, Wallingford: Triangle, pp.11–36

Hanna, G., Kundiger, E. and Larouche, C. (1990) 'Mathematical achievement of grade 12 girls in fifteen countries' in Burton, L. (1990) pp.87–97.

Harris, D. quoted in Welton, M. (1991) 'The life and times of adult education in North American liberal democracy' *International journal of University Adult Education*, 30(1), pp.13–38.

Harris, J. (1987) 'Australian Aboriginal and Islander Mathematics' *Australian Aboriginal Studies*, 2.

Harris, M. (1991) *School, Mathematics and Work*, Basingstoke: Falmer Press.

Harris, M. (1995a) 'Finding common threads: researching the mathematics in traditionally female work' in Coben, D. (1995a), pp.18–23.

Harris, M. (1995b) 'Mathematics in women's work: making it visible' in Coben, D. (1995b), pp.45–48.

Harris, M. (ed) (1991) *School, Mathematics and Work*, Basingstoke: Falmer Press.

Harris, M. and Evans, J. (1991) 'Mathematics and workplace research' in Harris, M. (1991) pp.123–131.

Hawking, S. (1988) *A Brief History of Time*, London: Bantam Press.

Head, D. (1991) 'Education at the bottom' in Westwood, S. and Thomas, J. (1991) pp.70–91.

Hedoux, J. (1982) 'Des publics et des non-publics de la formation d'adults: Sallaumines-Noyelles-sous-Lens des 1972' *Revue Francaise Sociologique* Avril-Juin, pp.253–274.

Henderson, D. (1990) 'The masquerade of formal mathematics and how it damages the human spirit' in Noss, R. *et al* (1990) pp.115–118.

Her Majesty's Inspectorate (1985) *Mathematics from 5 to 16*, London: HMSO

Hersh, R. (1986) 'Some proposals for reviving the philosophy of mathematics', in Tymoczko, T. (ed) *New Directions in the Philosophy of Mathematics*, Boston: Birkhauser.

Highet, G. (1991) 'Gender and education: a study of the ideology and practice of community based women's education' in Westwood, S. and Thomas, J.E. (eds) *The Politics of Adult Education*, Leicester: NIACE.

Hilton, P. (1980) 'Maths anxiety: some suggested causes and cures; part 1' *Two-year College Mathematics Journal*, 11(3), pp.174–188.

Hind, G (1993a) *Adult Numeracy Working Paper 93–4*, Department of Mathematics, University of Essex.

Hind, G (1993b) *Responsible Citizenship Working Paper 93–3*, Department of Mathematics, University of Essex.

Hind, G (1993c) *Figure Work, Working Paper 93–2*, Department of Mathematics, University of Essex.

Hind, G. (1995) 'The role of informal learning in adult numeracy' in Coben, D. (1995), pp.71–84.

Hoar, M. et al (eds) (1995) *Life Histories and Learning: Language, the Self and Education.* Centre for Continuing Education, University of Sussex and the School of Continuing Education, University of Kent.

Holfstadter, D. (1982) *Scientific American*, 246(5), pp.16–23.

Houle, C. (1960) 'The education of adult educational leaders' in Knowles, M. (ed) *Handbook of Adult Education in the United States*, Washington: Adult Education Association of the USA.

Huff, D. (1985) *How to Lie with Statistics*, Harmondsworth: Penguin.

Isaacson, Z. (1989) Of Course You *Could* Be an Engineer, Dear, But Wouldn't You *Rather* Be a Nurse or Teacher or Secretary? in P. Ernest (1989) pp.188–194

Isaacson, Z. (1991) 'The marginalisation of girls in mathematics: some causes and some remedies' in Pimm, D. (1991) pp.95–108.

Jackson, K. (1981) quoted in Thompson, J. (1981) p.104.

Jarvis, P. (1983) *Adult and Continuing Education: Theory and Practice* Beckenham: Croom Helm.

Jaworski, B. (1991) '"Is" versus "seeing as": constructivism and the mathematics classroom' in Pimm, D. (1991.) pp.287–296.

Johnston, R. (1993) 'Praxis for the powerless and punchdrunk' *Adults Learning*, 4(6), pp.146–148.

Joseph, G.G (1990) 'The politics of anti-racist mathematics' in R. Noss *et al* (1990) pp.134–142

Joseph, G.G. (1987) 'Foundations of Eurocentrism in mathematics' *Race and Class*, 28(3).

Joseph, G.G. (1991) *Non-Western Roots of Mathematics*, London: I.B. Tauris.

Jowitt, J. (1995) 'Ethnic minorities and continuing education', in *The Proceedings of the European Research Conference Adult Learning and Social Participation: The Changing Research Agenda*, Strobl, Austria: ESREA.

Kaisser, G. (1996) 'Equity in mathematics education: description of the current debate' a paper given at to The 8th International Conference on Mathematical Education, Seville, Spain.

Kane, R.B. (1970) 'Readability of mathematics textbooks revisited' *The Mathematics Teacher*, 63, pp.579–581.

Kant, I.(1783) *Prolegomena to any future Metaphysics*

Keddie, N. (1981) 'Adult education: an ideology of individualism' in Thompson, J. (1981) pp.45–64.

Keitel, C. (1996) 'Teaching maths anxiety: a circle of aversion to mathematics with teachers and students' a paper given to The 8th International Conference on Mathematical Education, Seville, Spain.

Kuhn, T, S. (1970) *The Structure of Scientific Revolutions*, Chicago: University of Chicago Press.

Knijnic, G. (1996) 'Adult numeracy and the interrelations between academic and popular knowledge' a paper given at ALM3 Conference, Brighton, England.

Knowles, M. (1970) *The Modern Practice of Adult Education: Andragogy v Pedagogy*, Surrey: Association Press.

Knowles, M.S. (1972) 'Self directing enquiry: innovations in teaching styles and approaches based upon adult learning', *Journal of Education for Social Work*, 8 (2).

Knowles, M.S. (1975) *Self-directed Learning: A Guide for Learners and Teachers*, New York: Association Press.

Knowles, M.S. (1980) (2nd Ed) *The Modern Practice of Adult Education from Pedagogy to Andragogy*, Chicago: Association Press.

Knowles, M.S. (1984) *The Adult Learner: a Neglected Species*, Chicago: Follet.

Koestler, A (1959) *The Sleepwalkers: A History of Man's Changing Vision of the Universe*, New York: Macmillan.

Kress, G. (1985) 'Socio-linguistic development and the mature language user: different voices for different occasions' in Wells, G. and Nicholls, J. (eds) *Language and Learning: An international Perspective*, London: Falmer Press.

Ladson-Billings, G. (1995) 'Making mathematics meaningful in multcultural contexts' in Secada, W., Fennema, E. and Adajian, L. (eds) *New Directions for Equity in Mathematics Education*, Cambridge: Cambridge University Press, pp. 126–145.

Lakatos, I. (1976) *Proofs and Refutations*, Cambridge: Cambridge University Press.

Lave, J. (1988) *Cognition in Practice*, Cambridge: Cambridge University Press

Lea, M. and West, L. (1994) 'Identity, the adult learner and institutional change' in Armstrong, P. *et al* (1994) (eds.) *Reflecting on changing practices, contexts and identities*, Hull: SCUTREA pp.75–77.

Leder, G. (1990) 'Gender and Classroom Practice' in Burton, L. (1990), pp.9–19.

Lee, A. (1995) 'Discourse, mathematics and numeracy teaching' *Numeracy in Focus*, 1, pp.47–51.

Leicester, M. (1993) *Race for a Change in Continuing and Higher Education*, Buckingham: SRHE and Open University Press.

Leonelli, E. and Schwendeman, R. (1994) (eds) *The Massachusetts Adult Basic Education Math Standards*, Massachusetts: The ABE Math Standards Project.

Lerman, S. (1992) 'Learning mathematics as a revolutionary activity' in Nickson, M. and Lerman, S. (eds) (1992) pp.170–179

Lesne, M. (1985) 'The training of adult educators' in Debesse, M. and Mialaret, G. (eds) *Educational Sciences vol. 8*, Athens: Diptyho.

Llorente, J.C. (1996) *Problem Solving and Constitution of Knowledge at Work*, Department of Education, University of Helsinki.

Lovejoy, F. and Barboza, E. (1984) 'Feminine mathematics anxiety: a culture-specific phenomenon' in Burns, R. and Sheehan, B. (eds) *Women and Education*, La Trobe University, Victoria.

Lovett, T (ed) (1988) *Radical Approaches to Adult Education: A Reader*, London and New York: Routledge.

Lovett, T. (1975) *Adult Education, Community Development and the Working Class*, London: Ward Lock Educational.

Lovett, T. (1988) 'Community Education and Community Action' in Lovett, T (1988) pp.141–163.

Lyotard, J. (1984) *The Postmodern Condition: A Report on Knowledge*, Manchester: Manchester University Press.

Maher, C. (1996) 'Critical mathematics education' a paper given to The 8th International Conference on Mathematical Education, Seville, Spain.

Maier, E. (1991) 'Folk Mathematics' in Harris, M. (1991), pp.62–66.

Malcolm, J. (1995) 'The competence of worker bees: the implications of competence-based education for educators of adults' in Collins, M. (1995) pp.61–74.

Malcolm, J. (1992) 'The culture of difference: women's education re-examined' in Miller, N. and West, L. (1992) pp.52–55.

Marr, B. and Helme, S. (1991) *Breaking the Maths Barrier*, Melbourne: DEET.

Marshall, J. (1989) 'Foucault and Education' *Australian Journal of Education*, 33(2), pp.99–113.

Maxwell, J. (1988) 'Hidden messages' in Pimm, D. (1988) pp.118–121.

Maxwell, J. (1988) 'Mathephobia' in Ernest, P. (1988) *The Social Context of Mathematics Teaching*, School of Education, University of Exeter, pp.26–33.

Maxwell, J. (1991) 'Hidden messages' in Harris, M. (1991) pp.67–70.

Mayo, P. (1994). 'Synthesizing Gramsci and Freire: possibilities for a theory of radical adult education', *International Journal of Lifelong Education*, 112: 2, pp.125–148.

McCaffery, J. (1993) 'Gender planning', *Adults Learnıng*, 5(3), pp.79–81.

McConnell, D. (1995) 'From distance learning to computer supported co-operative learning-a new paradigm for distance learning' in Bryant, I. (1995) pp.124–129.

McGivney, V. (1990) *Access to Education for Non-Participant Adults*, Leicester. NIACE.

McPeck, J. (1981) quoted in Abraham, J. and Bibby, N. (1992)

Meighan, R. (1986) *A Sociology of Education*, Eastbourne: Holt, Rinehart and Winston.

Mellin-Olsen, S. (1987) *The Politics of Mathematics Education*, Dordrecht: Reidal.

Mellin-Olsen, S. (1990) 'Liberation of knowledge' in R.Noss *et al* (1990) pp.173–187

Mezirow, J. (1977) 'Perspective transformation' *Studies in the Education of Adults*, 9(2).

Mezirow, J. (1981) 'A critical theory of adult learning and education' in Tight, M. (ed) *Adult Learning and Education*, London: Routledge, pp.124–140.

Mezirow, J. (1994) 'Understanding transformation theory' *Adult Education Quarterly*, 44(4), pp.222–232.

Mezirow, J. (ed) (1990) *Fostering Critical Reflection in Adulthood*, San Francisco: Jossey-Bass.

Miller, N. and West, L. (eds) (1992) *Changing Culture and Adult Learning*, Boston: SCUTREA.

Ministry of Reconstruction (1919) *The Final Report of the Adult Education Committee*, Cmd.321, London: HMSO.

Modood, T. (1992) *Not Easy Being British: Colour, Culture and Citizenship*, Stoke-on-Trent: Trentham Books.

Munn, P. and MacDonald, C. (1988) *Adult Participation in Education and Training*, SCRE.

National Institute of Adult Education (1970) *Adequacy of Provision*, NIAE.

National Science Foundation (1983) *Educating Americans for the 21st Century*, Washington, DC. pp.13–14.

NCC (1990) *Core Skills 16–19: A Response to the Secretary of State* York: NCC.

Nesbitt, T. (1995) 'Lost in space: mathematics education for adults', Paper given to the 1995 AERC Conference.

Newman, M. (1994) *Defining the Enemy: Adult Education in Social Action*, Sydney: Stewart Victor Publishing.

Newsome Report (1983) *Half Our Future*, London: HMSO

Neyland (1995) 'Teacher's Knowledge' in Almeida, D. and Ernest, P. *Perspectives on Maths*, Exeter: University of Exeter Perspectives 53, pp.80–102.

NIACE (1989) *Adults in Higher Education: A Policy Discussion Paper*, Leicester: NIACE.

NIACE Replan (1991) *Women Learning: Ideas, Approaches and Practical Support*, Leicester: NIACE.

Nickson, M. (1992) 'Towards a multi-cultural mathematics curriculum' in Nickson, M. and Lerman, S. (1992) pp.128–135.

Nickson, M. and Lerman, S. (eds) (1992) *The Social Context of Mathematics Education*, London: Southbank Press.

Northedge, A. (1994) 'Access as initiation into academic discourse' in Lemelin, R. (comp) *Issues in Access to Higher Education*, Portland: University of Southern Maine, pp.145–151.

Noss, R. (1990) 'The National Curriculum and mathematics: political perspectives and implications' in Nickson, M. and Lerman, S. (eds) pp.62–64.

Noss, R. *et al* (eds) (1990) *Political Dimensions of Mathematics Education: Action and Critique*, London: Department of Mathematics, Statistics and Computing, Institute of Education, University of London.

Noss, R. (1991) 'The computer as a cultural influence in mathematical learning' in Harris, M. (1991) pp.77–92.

NSDC (1995) *Adult Numeracy Teaching: Making Meaning in Mathematics*, Melbourne: National Staff Development Committee for Vocational Education and Training.

Oats, T. (1990) *Developing and Piloting the NCVQ Core Skills Units* Report 16. London: NCVQ.

Pai, Y. (1990) 'Cultural pluralism, democracy and multicultural education', in Cassara, B. B. (ed) *Adult Education in a Multicultural Society*, London and New York: Routledge.

Paulos, J. (1990) *Innumeracy*, London: Penguin.

Payne, J. (1992) 'The adult education centre: cultures of affluence, cultures of poverty' *International Journal of Lifelong Education*, 11(3), pp.217–233.

Percy, K.(1995) 'Research into adult self-directed learning in Britain and its implications for educating the adult educator' in Collins, M. (1995) pp.213–219.

Percy, K., Barnes, B., Machell, J. and Graddon, A. (1988) *Learning in voluntary organisation,* Leicester: Unit for the Development of Adult Education.

Pimm, D. (ed) (1991) *Mathematics, Teachers and Children*, London: Hodder and Stoughton and the Open University Press.

Pimm, D.(1990) 'Mathematical versus political awareness' in R. Noss *et al* (1990) pp.200–204.

Plaza, P. (1996) Poster Session: Materials from the Adult School in Madrid, The 8th International Conference on Mathematics Education, Seville, Spain.

Plowden Report (1967) *Children and their primary schools*, London: HMSO.

Preece, J. (1995) 'Discourse and culture' in Bryant, I. (1995) pp.154–159.

Preece, J. (1996) 'Historicy repeats itself: power-knowledge games in continuing education', paper given to ESREA Access to Higher Education Conference, University of Leeds.

Presmeg, N. (1988) 'School mathematics in culture-conflict situations' *Educational Studies in Mathematics*, 19, pp.163–177.

Radical Statistics (1995) *Newsletter 61 Winter 95*, London: Radical Statistics.

Rees and Barr (1984) *Diagnosis and Prescription*, Harper and Row quoted in Wareham (1993).

Resnick, L. (1986) 'The development of mathematical intuition' in Perlmutter, M. (ed) *Minnesota Symposium on Child Psychology*, 19, Hillsdale NJ: Erlbaum.

Robbins Report (1963) *Committee on Higher Education Report*, CMND.2154, London: HMSO.

Roche, M. (1992) *Rethinking Citizenship*, Polity Press.

Rockhill, K. (1995) 'Challenging the inclusive masks of adult education' in Collins, M. (1995) pp.1–6.

Rodgers, M. (1990) 'Mathematics: pleasure or pain?' in Burton, L. (ed) (1990) pp.29–37.

Rogers, A. (1993) ' Adult Learning Maps and the Teaching Process' *Studies in the Education of Adults*. 25(2) pp.199–220.

Rogers, C. (1967) *On becoming a person: a therapist's view of psychotherapy*, Constable and Co.

Rogers, C. (1969) *Freedom to Learn*, Ohio: Merrill Publishing Company.

Russell Report (1973) *Adult Education: A Plan for Development* London: HMSO.

Ruthven, K. (1987) 'Ability Stereotyping in Mathematics' *Educational Studies in Mathematics*, 18, pp.243–253.

Sargant, N. (1990) *Learning and Leisure: A Study of Adult Participation in Learning and its Policy Implications*, Leicester: NIACE.

Sargant, N. (1993) *Learning for a Purpose*, Leicester: NIACE.

Saxe, G. (1988) 'Candy selling and math learning' *Educational Researcher*, (August-September) pp.185–221.

Schon, D. (1982) *The Reflective Practitioner: How Professionals Think in Practice*, New York: Basic Books.

Schon, D. (1987) *Educating the Reflective Practitioner*, London: Jossey-Bass.

Schuller, T. (1978) *Education Through Life*, Fabian Society.

Scott-Hodgetts, R. (1986) 'Girls and mathematics: the negative implications of success' in Burton, L. (1986).

Scott-Hodgetts, R. (1992) 'The national curriculum: implications for the sociology of the mathematics classroom' in Nickson, M. and Lerman, S. (1992) pp.10–25.

Scribner, S. (1984) 'Cognitive studies of work' *The Quarterly Newsletter of the Laboratory of Comparative Human Cognition*, 6, pp.1–49.

Scribner, S. and Cole, M. (1973) 'Cognitive consequences of formal and informal education' *Science*, 182 pp.553–559.

Seidler, V.J. (1994) *Unreasonable Men, Masculinity and Social Theory*, London: Routledge.

Sellars, J. (1995) 'Quilting a life history', in Hoar, M. et al (1995) pp.136–140.

Shuard, H. (1982) 'Differences in mathematical performance between girls and boys' in Cockcroft, W. (1982).

Shuard, H. (1986) 'The relative attainment of girls and boys in mathematics in the primary years' in Burton, L. (1986).

Siraj-Blatchford, I. (1990)' Access to what? Black students, perceptions of initial teacher education', *Journal of Access Studies*, 5(2), pp.177–187.

Skelton, C. (1985) 'Gender Issues in a PGCE Teacher Training Programme' Unpublished MA Thesis, Education Department, University of York.

Skelton, C. and Hanson, J. (1989) 'Schooling the teachers: gender and initial teacher education' in Acker, S. (ed) *Teachers, Gender and Careers*, London: Falmer Press, pp.109–122.

Smith, D. and Tomlinson, S. (1989) *The School Effect: The Study of Mult-racial Comprehensives*, Lancaster Policy Studies Institute: University of Lancaster.

Snow, C.P. (1959) *The Two Cultures and the Scientific Revolution*, London: Cambridge University Press.

Spencer, S. (1996) 'Asking open ended questions in adult numeracy classes?' *Mathematics Support Newsletter*, 4&5, p.21.

Spendiff, A, (1987) *Maps and Models: Moving Forward with Feminism*, Breaking Our Silence Series, London: WEA.

Stanworth, M. (1983) *Gender and schooling: a study of sexual divisions in the classroom*, London: Hutchinson.

Steele, T. (1993) 'Taking the epistemology:what happened to that discreet object of knowledge?' in Miller, N. and Jones, D. (eds) *Research reflecting practice*, Manchester: Standing Conference of University Teaching and Research in the Education of Adults, pp.35–37.

Strong, M. (1977) *The Autonomous Adult Learner*, Unpublished M.Ed. dissertation, Nottingham: University of Nottingham.

Stuart, M. (1995) 'Accrediting selves: recognition of prior learning, the self and the academy' in Hoar, M. *et al* (1995) pp.147–150.

Stuart, M. (1995) 'If experience counts then why am I bothering to come here?: AP(E)L and learning' in Stuart, M. and Thompson, A. (1995) (eds) *Engaging With Difference*, Leicester: NIACE, pp.158–170.

Swann Report (1985) *Education for All*, London: HMSO

Taylor, R. (1986) 'Problems of inequality: the nature of adult education in Britain' in Ward, K. and Taylor, R. (eds) (1986) *Adult Education and the Working Class: Education for the Missing Millions*, London: Croom Helm pp. 1–26.

Taylor, R. (1990) 'Racism, ethnicity and university adult education', *Studies in the Education of Adults*, 22(2), pp.211–220.

Taylor, R., Rockhill, K. and Fieldhouse, R. (1985) *University Adult Education in England and the USA*, Beckenham: Croom Helm

Tennant, M. (1994) 'Response to understanding transformation theory' *Adult Education Quarterly*, 44(4), pp.233–235.

Thomas, E. (1995) 'Universities and the community: whose rights? whose duties?' in Collins, M. (1995) pp.35–40.

Thomas, J. E. and Harries-Jenkins, G. (1975) 'Adult Education and Social Change' *Studies in Adult Education*, 7(1), pp.1–15.

Thompson, J.(1981) *Adult Education for a Change*, London: Hutchinson

Thorstad, I. (1992) 'Adult Numeracy and Responsible Citizenship' *Adults Learning*, 4(4), pp.104–105.

Tobias, S. (1978) *Overcoming Math Anxiety*, Boston: Houghton Mifflin.

Tough, A. (1971) 'The Adults Learning Projects', in *Research in Education Series No. 1*, Toronto: Ontario Institute for Studies in Education.

Tuckett, A. (1991) 'Counting the cost: managerialism, the market and the education of adults in the 1980s and beyond' in Westwood, S. and Thomas, J. (1991) pp.21–43.

Uden, T. (1996) *Widening participation*, Leicester: NIACE.

University Council for Adult Continuing Education (1990) *Report of the Working Party on Continuing Education Provision for the Minority Ethnic Communities, Occasional Paper No. 2*, Warwick: UCACE.

Usher, R. (1989) 'Locating adult education in the practical' in Bright, B. (ed) (1989), pp.65–93.

Usher, R. (1995) 'Telling the story of self/deconstructing the self of the story' in Bryant, I. (1995) pp.178–183.

Vergnaud, G. (1983) 'Multiplicative structures' in Lesh, R. and Landau, M. (eds) *Number Concepts and Operations in the Middle Grades*, New York: Academic Press, pp.127–174.

Vergnaud, G. (1988) 'Multiplicative structures' in Hiebert, M. and Behr, J. (eds) *Number Concepts and Operations in the Middle Grades*, National Council for Teachers of Mathematics.

Verhage, H. (1990) 'Curriculum development and gender' in Burton, L. (1990) pp.60–71.

Volmink, J. (1990) 'The constructivist foundation of ethnomathematics' in R.Noss *et al* (1990) pp.200–204

Vygotsky, L. (1962) *Thought and Language*, Cambridge: MIT Press.

Walker, J. (1988) *Louts and Legends*, Sydney: Allen and Unwin quoted in Watson, I. (1993) 'Education, class and culture: the Birmingham ethnographic tradition and the problems of the new middle class' *British Journal of Sociology of Education*, 14(2) pp.179–197.

Walkerdine, V. (1983) 'It's only natural: rethinking child-centred pedagogy' in Wolpe, A. and Donalds, J. (eds.) *Is anyone here from education?*, London: Pluto Press.

Walkerdine, V. (1988) *The Mastery of Reason*, London and New York: Routledge

Walkerdine, V. (1989) *Counting Girls Out*, London: Virago Press

Wareham, S. (1993) 'Reading comprehension in written mathematical problems' ALBSU *Viewpoints 16*, London: ALBSU.

Weil, S. (1993) ' Access: Towards education or miseducation? Adults imagine the future' in Thorpe, M. *et al* (eds.) *Culture and Processes of Adult Learning*, London: Routledge.

Weiler, K. (1995) 'Freire and a feminist pedagogy of difference' in Holland, J and Blair, M. with Sheldon, S. (eds.) *Debates and Issues in Feminist Research and Pedagogy*, Multilingual Matters and the Open University, pp.23–44.

West, L. (1996) 'Access to what and on whose terms?' paper given to ESREA Access to Higher Education Conference, University of Leeds.

Westwood, S. (1981) 'Adult education and the sociology of education: an exploration' in Thompson, J. (1981) pp.31–34.

Westwood, S. (1992) 'Power/knowledge: the politics of transformative research' *Studies in the Education of Adults*, 24(2) pp.191–198.

Westwood, S. and Thomas, J. (eds) (1991) *The Politics of Adult Education*, Leicester: NIACE.

Williams, R. (1961) *The Long Revolution*, Harmondsworth: Penguin.

Williams, R. (1976) *Keywords* Fontana/Croom Helm .

Willis, S. (1989) *Real girls don't do maths: gender and the construction of privilege*, Victoria: Deakin University Press

Willis, S. (1995) 'Gender justice and the mathematics curriculum: four perspectives' in Parker, L. Rennie, L. and Fraser, B. *Gender, Science and Mathematics: Shortening the Shadow*, Kluwer Academic Publishers.

Willis, S. (1996) 'Perspectives on social justice, disadvantage and the mathematics curriculum: a view from Australia' a paper given at ICME-8, Seville.

Willis, S. (ed) (1990) *Being Numerate: What Counts?*, Melbourne: ACER.

Wilson, A. (1995) Keynote Address *Conference Proceedings of University Continuing Education with the Minority Ethnic Communities*, Leeds: UACE.

Wiltshire, H. (1987) quoted in *The WEA and the Black Communities*, WEA.

Withnall, A. (1995) *Older Adults' Needs and Usage of Numerical Skills in Everyday Life*, Lancaster University

Wittgenstein, L. (1956) *Remarks on the Foundations of Mathematics*, Oxford: Blackwell

Wolffensperger, J. (1993) 'Science is truely a male world'. the interconnectedness of knowledge, gender and power within university education' *Gender and Education*, 5(1) pp.37–54

Woodrow, D. (1989) 'Multi-cultural and anti-racist mathematics teaching' in P. Ernest (1989) pp.229–235.

Yasukawa, K. and Johnston, B. (1994) 'A numeracy manifesto for engineers, primary teachers, historians ... a civil society – can we call it theory?' *Proceedings of the Australian Bridging Mathematics Network Conference*, Sydney University, pp.191–199.

Zaslavsky, C. (1990) 'Effects of Race and Class on Mathematics Education in the United States' in R. Noss *et al* (1990) pp.256–263

Zeldin, D. (1992) 'Cultures and double beings: linking adult learners and their environments' in Miller, N. and West, L (1992) pp.78–81.

Index

Printed in the United Kingdom
by Lightning Source UK Ltd.
102657UKS00001B/247-387

9 781862 010079